Not The Red Baron

A biography by

Geoff Pridmore

Chiselbury

First published 2014
Second edition Copyright © 2023 Geoff Pridmore

Published by Chiselbury Publishing, a division of Woodstock Leasor Limited
14 Devonia Road, London N1 8JH, United Kingdom

www.chiselbury.com

ISBN: 978-1-916556-03-4

The moral right of Geoff Pridmore to be identified as the author of this work is asserted.

This is copyright material and must not be copied, reproduced, transferred, distributed, leased, licensed or publicly performed or used in any way except as specifically permitted by the publishers, as allowed under the terms and conditions under which it was purchased or as strictly permitted by applicable copyright law. Any unauthorised distribution or use of this text may be a direct infringement of the publisher's rights and those responsible may be liable in law accordingly.

Chiselbury Publishing hereby exclude all liability to the extent permitted by law for any errors or omissions in this book and for any loss, damage or expense (whether direct or indirect) suffered by any third party relying on any information contained in this book.

This book is dedicated to all those wonderful people who fly to entertain.

Table of Contents

Chapter 1 – Last Flight 1
Chapter 2 - Doodlebug Summer 20
Chapter 3 - A New Life in Devon (1961-1967) 29
Chapter 4 - By the Way, I've bought an aeroplane … 37
Chapter 5 – And There it Lay 63
Chapter 6 – Airymouse Flies Again 83
Chapter 7 – Doing Business in Ermington 100
Chapter 8 – Ireland: The Ulster Air Show and Air Spectacular 130
Chapter 9 – Barnstorming & Test Flying 148
Chapter 10 – Inquest 220
Chapter 11 – Into the Sunset 264
Memories 282
About the Author 298
The Story behind "Not the Red Baron" 299

Chapter 1 – Last Flight

Many returned home, some stayed forever, none will be forgotten.

(Inscription on a memorial stone dedicated to those aircrews that served at Dunkeswell aerodrome during WW2.)

Dunkeswell aerodrome stands on high ground in east Devon, just over three miles south of the Somerset county boundary. From the completion of its construction in 1943 to the end of the war, the airfield was home to both the United States Army Air Force and later the United States Navy, both services operating squadrons of Liberator bombers from the base. Today, the taxiways circumnavigating the perimeter are potholed and a trading estate has blossomed like a poppy from a bloody field to the west of the airfield. Unlike many other former WW2 aerodromes, Dunkeswell lives on as an airfield, home to pleasure flyers, private companies and The Devon School of Flying. And, throughout the eighties and early nineties, it was a sometime base for an aviator whose growing fame as a display pilot was beginning to spread across Britain and continental Europe; flying a full scale replica Fokker Dr.1 triplane in the guise of the Red Baron; his name - Robin Bowes.

From this airfield on a hot July evening in 1995, Robin would make a second flight to the majestic ornamental gardens of Stourhead in Wiltshire where he would rendezvous with three other display pilots and perform with them an aerial battle of the type that would originally have been fought, and indeed witnessed, by a generation of men and women almost all, save a precious few, gone forever. His fellow display pilots, Doug Gregory, Des Biggs and Chris Mann, flying replica British SE5a biplanes were due to rendezvous with him at 8 pm at over 500 feet above the gardens. Flying vicious circles above the spectators, the four aeroplanes were to re-enact a WW1 dogfight. Simulating mock combat, they would engage his scarlet triplane in a battle that was to climax in the Red Baron's demise. At the conclusion of battle, emitting smoke from a special canister fitted to the triplane, Robin would bank away beyond view of the crowds in the wooded gardens below, to fly safely home. It was a battle he and his friends had fought many times before across British skies; he, as the Baron, always destined to lose, history dictating that there never could be any other eventual outcome for that had been von Richthofen's fate and so each night it was to be Robin's fate too. Only this night, fate had its own rendezvous to keep with the man who flew as the Red Baron.

Battle in the Sky

The battle in the sky was by no means an accurate representation of history. Air ace, Baron Manfred von Richthofen, flew the Fokker triplane for a mere six months out of an 18 month career of aerial combat above the trenches in which, of his 80 confirmed victories, only 19 were claimed through the twin Spandau gun barrels of the triplane; though the type was to become synonymous with his name and his death; his fate sealed not by SE5a fighters, but by two Sopwith Camels and an Australian Field Battery. On the morning of Sunday April 21, 1918, recklessly in pursuit of a British

Sopwith Camel biplane, and himself pursued by another Camel piloted by Canadian RFC Captain A.R. Brown, Freiherr Manfred von Richthofen chased his quarry two miles behind British lines flying dangerously low at between 60 to 100 feet and well within range of Aussie gunners Cedric Popkin, William Evans and Robert Buie. Caught in small arms crossfire, the Red Baron lost control of his triplane crashing into a field of beet alongside the Bray-Corbie road. Germany's highest scoring air ace was found dead at the scene still strapped into his cockpit. No post mortem as such was carried out yet two hastily performed medical examinations showed a bullet wound had entered his torso from in front of the right arm pit and exited half an inch above his left nipple having been deflected after striking the spinal column. General Sir Henry Rawlinson, commanding the British Fourth Army, accepted the possibility that the gunners had indeed caught the Red Baron in crossfire thereby causing his downfall. The newly formed Royal Air Force never accepted his conclusion insisting Captain Brown had mortally wounded the Baron in the pursuit. Richthofen's mother, Baroness Kunigunde, also never accepted the possibility that her noble son had been brought down by common artillerymen. Whoever pulled the trigger, one thing is for sure, upon his death Baron von Richthofen, Prussian aristocrat and air ace extraordinaire, became legend.

Richtofen, as Geschwader Commander of Jasta 1, had admired the innovatory 'dri-dekker' type aeroplane well enough to encourage aircraft designer Anthony Fokker to derive a German equivalent from the original British Sopwith triplane, which had proved highly successful for a short while. The Germans were lagging behind in the air technology war and desperately needed a machine that would equal the RFC's best. But no sooner had Fokker built his Dr1, than the type was abandoned by the British in favour of the new Sopwith Camel and Spad 13. In a bid to keep pace, the Germans simply found themselves trying out an experimental type that was to claim the lives of some of their best flyers in accidents and cost the war effort heavily in money and time. Before its brief life was over, however, the triplane as war bird did prove suitable to a few of Germany's best pilots among them Werner Voss whose incredible skills, and undoubted good luck, found the triplane a highly manoeuvrable killing machine if handled correctly. Eventually, it was Werner Voss who met his end in combat with SE5a aircraft - not von Richthofen. And maybe it should have been his life and death commemorated in mock battle that night at Stourhead; but such is the making of myth and legends.

One Man's Life cut short

The flight to Stourhead, the span of a county away, would take approximately thirty-five minutes flying at some eighty knots – almost half the time as the actual drive from the village of Ermington where Robin lived with his partner, Sheila Truscott. Ermington was over fifty miles south west of Dunkeswell and though both parishes share the ancient county of Devon, both are different in certain respects. Dunkeswell, its oldest farms and cottages – many thatched – nestle in the wooded coombe below the aerodrome, together with the pub and church of St. Nicholas in which is contained a plaque to the names of the airmen who didn't return and a memorial organ donated by the US Navy. Above on the hilltop is the post war development of houses and a garage in an area where the Navy crews were once billeted. Among those names on the plaque, is that of one pilot in particular – Joseph Patrick Kennedy Junior.

The man that might have been …

Eldest of the famous Kennedy clan, it is reasonable to speculate that it might have been Jo Kennedy running for office in 1960, not his younger brother, John. Had he not volunteered to fly a seemingly suicidal mission, one can only speculate as to how different a course post-war history might have run: The Bay of Pigs? Vietnam? The Cuba Missile crisis? The race to the moon? Civil rights? Dallas?

In his father's eyes, Joe Junior was the natural and true heir to the presidency of the United States. His maternal grandfather, John F. Fitzgerald, known by all as 'Honey Fitz', and the first son of emigrant parents to become Mayor of Boston, cradled Jo Junior as a baby and prophesied that the infant would grow up to become president of the United States.

History records little of Joe's short life, yet the significance to world history had he survived his mission, should not be underestimated. Shortly before his death, he was best man at younger sister Kathleen's (aka 'Kick') wedding to English aristocrat William ('Billy') John Robert Cavendish, Marquis of Hartingdon. At the wedding, both best man and groom attended in their respective uniforms of US Navy flier and Coldstream Guards officer. Within four months both men had been killed in action; Kathleen would die in a plane crash in 1948.

Jo Junior's death is even more tragic considering that he had volunteered for a 'suicide' mission – something he could have walked away from. Privilege of connections and family background would have seen to that, but circumstances in the Pacific not long before had already set in motion Jo's destiny with fate. Twelve months earlier, in the early hours of August 2nd, Jo's younger brother, USNR Lieutenant John F. Kennedy, was the commanding officer of PT109, a small, but fast torpedo boat operating off Tulagi and the Russell Islands in the South Pacific, when a Japanese destroyer rammed the boat slicing it in two. A hastily shouted warning from a crewmember came too late as the giant bows of the destroyer rose out of the darkness and hit PT109 amidships. The black sky, which had veiled the destroyer, was now illuminated by a ball of fire as the twin fuel tanks of the boat ignited on impact. Torpedo boats owed their speed to the aviation engines that powered them, but with highly inflammable aviation fuel tanks onboard such an impact could be catastrophic. The explosion witnessed from other US Navy boats in the flotilla, convinced those watching there could be no survivors and therefore a search was not conducted. But men had survived.

Lt. Kennedy, thrown backward onto the surface of the water, aggravating childhood back problems, counted ten of his 12 man crew floating in the water, all like him in varying degrees of pain and shock. Survival now was in the shape of PT109's front section, which had stayed afloat.

By daylight the survivors abandoned the rapidly sinking bows in an effort to swim to an island they estimated to be some three miles away, and free of Japanese occupation. Kennedy, despite his injured back, towed the boat's engineer, Pat MacMahon, now semi-conscious and badly burned, by gripping the crewman's life belt strap between his teeth. The years as a boy spent swimming in the family pool and in the waters off Hyannis Port, Massachusetts, now, unexpectedly, were paying off.

Five hours later they reached the island. Kennedy pulled MacMahon ashore, counted the survivors and though exhausted, waded back into the water to swim out in an effort to find a passing friendly boat. All through the night he swam only to return to the island in the morning where he collapsed. Six days later, Lt. Kennedy and the crew of PT109 were rescued.

Admiral Halsey awarded Kennedy the Navy and Marine Corps medal: "For extremely heroic conduct as commanding officer of Motor Torpedo Boat 109." He was also awarded the Purple Heart.

Joseph P. Kennedy Senior, keen to take advantage of the promotional opportunities provided by the story, made sure his son's heroics would make headlines across the US and the free world. However, the heir to the dynasty, Jo Junior, who had always enjoyed a loving but highly competitive relationship with his sibling, had been eclipsed. And in the highly competitive environment that was the Kennedy family, he wasn't about to allow Jack to hold centre stage alone.

Jo Junior had been studying law at Harvard in the autumn of 1941 when he led the family into the war effort by volunteering to join the US Navy as a trainee pilot. After flight training, his first operational sorties were patrols over the Caribbean and by September 1943, he was posted to Britain. As part of Fleet Air Wing 7, Jo's squadron replaced the 479th Anti-Submarine Group of the USAAF at Dunkeswell in March 1944 – making it the only US Navy base in Britain. It wasn't long before the Wing was suffering heavy casualties that included Jo's co-pilot and a number of close friends.

Never one to give up a fight, he persuaded his crew to continue beyond their designated number of missions at least until D-Day was completed. At the end of July, Jo and his crew were stood down having completed their tour, but news of brother Jack's heroics on the other side of the world had caused Jo to rethink his strategy. To redress the balance in his favour, the 29 year-old airman rashly volunteered for what was openly termed amongst Navy fliers, a 'suicide' mission. Part of a secret US Navy project code-named Anvil, his assignment involved flying a robot plane packed with explosives in an attack against German V1 rocket sites. 'Robot' aircraft were in effect full-scale radio controlled flying bombs. In theory, once the pilot had set the weapon on its course, he and his co-pilot could bale out over England leaving radio control to be directed from two 'mother' planes that would guide the robot or 'drone' plane to its target. For Jo, a veteran of coastal sorties, most of which were undertaken from Dunkeswell, it was a mission too far. On the 12th August 1944 – just thirteen days before Robin Austin Bowes was born in Blackpool, where his mother had been evacuated as a consequence of V1 attacks on London – Lt. Kennedy flew the first of the US Navy's Anvil operations. Piloting a converted B-24 Liberator that had been gutted of its interior fittings in order to be crammed with high explosives. The plane, minus its normal operational crew, was destined for a V3 Supergun base at Mimoyecques just inland of the French coast at Cap Gris Nez. Kennedy, together with his co-pilot Wilford Willy, took off from Fersfield, Norfolk with the white painted bomber he'd brought up from Dunkeswell. Wilford Willy had considerable experience of radio-controlled projects and had been at the forefront of Operation Anvil throughout its development. The Navy considered him an expert and therefore insisted that he replace Kennedy's regular co-pilot Ensign James Simpson.

There was method in the plan. Tests carried out in Florida earlier that year, had demonstrated the deadly effects of high explosives made from a cocktail of napalm and petrol detonating around concrete pillbox structures. The occupants – rats in respect of the tests – suffocated in the intense fire. If the radio control piloting of the drone were accurate enough, the ferocity of the explosion and resulting flames would trap and suffocate the technicians and scientists inside the launch area. But the tests had also highlighted the problems of controlling the drones once the crew had baled out. The US Navy's 'Anvil' operation was at that time running alongside the USAAF's own 'Aphrodite' operation, but it was the Navy that was later favoured because they had had some measure of success in radio controlled drone flying.

It was a warm summer's evening on August 12th – barely ten days after the first anniversary of his brother's heroic survival in the Pacific – when Jo Kennedy powered the flying bomb down the runway and gently climbed to 2000 feet above the Suffolk coast. For Wilford Willy, there was barely room aboard to stand. Astern of the drone, the two 'mother' ships that were to guide the robot to its target rendezvoused with

Kennedy at 2000 feet and cruised a respectful 500 to 1000 ft back. This part of the operation in itself had been tricky. A collision over East Anglia, however scattered the habitation below, would be the biggest single explosion in the war's history. The mother ships were already airborne and circling before falling in behind the Liberator. Take off and rendezvous had been the most hazardous part of the mission. The drone was packed with 21,170lbs of explosives filling its flight deck, command deck, forward and aft bomb bays, and even the nose wheel compartment. Had it failed to take off, the crash would have all but destroyed the aerodrome and the immediate vicinity. Airborne, Kennedy and Willy, would have breathed somewhat easier.

The mother ships were even equipped with a television from which they could monitor the progress of the drone once the crew had baled out. Though the picture was grey and features were barely distinguishable, it was nonetheless a remarkable advancement in warfare and one that would be perfected for use in the cruise missiles of the Gulf War some fifty years later. The three planes formed the central part of a much larger flight that included P-51 Mustangs flying shotgun at a greater altitude, and a Mosquito. A DC3 Dakota was also airborne to watch the landing of the two-crew members in the sea and keep them in sight until a boat arrived to pick them up.

The nearest airborne witnesses to the disaster were aboard the two mother 'ships' – as the American crews referred to their aircraft. Nervously expecting to see Kennedy and Willy leap from the drone, Bill Worthen heard Kennedy radio: "Spade flush", the code that indicated he was passing control over to the radio control of the mother ship and would therefore bale-out imminently. In an instant of hearing Kennedy's voice, a ball of fire engulfed the Liberator; the resultant blast pushed the following plane upwards at what seemed to the crew to be no less than a 90- degree angle. The mosquito was thrown onto its back by the resulting blast sustaining damage to the Plexiglas nose from incoming fragments. Injured by the debris, the observer, Lt. McCarthy, retreated to the cockpit and assisted the pilot in preparing for an emergency landing.

By the time Worthen's plane had levelled out, all that could be seen was a pall of smoke. Immediately they went into a circle pattern in a bid to spot debris and crew. Below them, 147 properties covering 60 square miles had sustained damage from the blast. The course that should have kept them over water had not been kept. The bodies of Jo Kennedy and Wilford Willy were never found, though strands of parachute chords and silk fragments were found tangled in hedgerows. The explosion was so great in its ferocity that no witness flying that day could testify to having seen anything so powerful in its destructive force. Of the wreckage found in the Blythburgh Fen area were just a few mangled bits of heavier steel engine and landing gear components and some rubber fragments from one of the tyres. No one on the ground was hurt thankfully, but there had been witnesses. Jo Kennedy and Wilford Willy had paid the ultimate price for a mission too far.

Just days before his last mission, Jo had typed what was to be his last letter home to younger brother Jack from the Norfolk base. The letter was cheery, though its author was obviously fed up with being confined to base in the depths of the Norfolk countryside. With good humour, he acknowledged the regular news from the family that contained much of his siblings praises (and current state of health) from both family and senior officers in the Navy, and that, "no", he wasn't likely to get married at this point. Lady Sykes – Virginia Gilliat – was a friend. When news reached the Kennedy compound in Hyannis Port, the family were just finishing lunch. Two priests with Navy commissions approached the front porch – both in uniform – and after consultation with Joseph and Rose Kennedy, the priests departed, leaving the younger members of the clan – brothers, sisters, and cousins – to speculate with growing concern as to the reason behind this unannounced, official visit. Their worst fears were confirmed when Joseph Senior addressed them in the Sun Room. With just a few words, he told them of the tragic news from England, whereupon the old man turned, left the room and climbed the stairs to his bedroom where he closed the door and wept. After his father had left the room, Jack, now suddenly the eldest of the sibling clan, calmed the others:

"Jo would not want us sitting here crying. He'd want us to go sailing. Teddy, Joey, get the sails; we're going to go sailing."

By the time Fleet Air Wing 7 departed Dunkeswell at the end of the war, they left behind 183 officers and men whose names are listed on a memorial plaque in the church. In the years since those dark days, and most especially during the eighties and nineties, the family and friends of those that didn't return, have crossed the Atlantic to pay them homage. The visitors' book bears witness to the many lives affected. And the church organ left by the donation of those crews still plays to the sound of English hymns, and probably always will. Joe Kennedy Junior was posthumously awarded both the Navy Cross and the Air medal. He was well remembered by all who met him at Dunkeswell as a friendly, kind and polite young man – just one of the lads who stayed awhile; drank ale in the pub and walked the muddy lanes along with his comrades. Jo's name also lived on in the commission after the war of US Navy destroyer 'Joseph P. Kennedy', which played its own vital role in the Cold War.

Almost two decades later, and speaking as President of the United States, Jack Kennedy paid this tribute to his elder brother: "Joe should not have pushed his luck, but he had completed more combat missions in heavy bombers than any other pilot of his rank in the US Navy. The odds were never better than 50-50. Joe never asked for better odds than that."

A Devonshire Landscape Below

Ermington, the larger of the two villages, is a conurbation of stone and tile roofed cottages centred around the 15th century church of St.Peter and St.Paul – noted mainly for its crooked stone spire, one of three in the Westcountry; an abnormality caused not by the devil, as folklore has it, but the weight of stone on unseasoned wood. The crooked spire lends its name to one of two pubs, and there is a combined village stores-cum-post office. Yet despite these components both villages were largely ignored by those heading west for holidays and by those promoters at the West Country Tourist Board whose job it was to promote in their glossy brochures any number of villages, towns and cities within the region that might hold a special interest for the visitor; whether it be for play or relaxation, historical or geographical, or the combination of all of these. To be fair, it could never be an easy task choosing from the many hundreds of parishes of special interest situated within the Westcountry peninsular. Five counties from Wiltshire to Cornwall – six if you included the short-lived Avon – came within the board's remit. Indeed, both Devon villages were unremarkable tourist destinations neither boasting famous burial places, authors cottages, clotted cream dairies, adventure theme parks, or fields of scientific interest. For those who lived in these parishes and loved them, the lack of fame was never bemoaned. This, however, could not be said of Stourhead – the jewel in the crown of the West Country Tourist Board and most especially the National Trust.

A Very English Garden

They weren't quite sure how things would work out over the next few days. After the display at Stourhead, Robin was to fly on to Rendcomb, Gloucestershire, where he would spend the night before piloting a Boeing Stearman biplane in display for Vic Norman's Cadbury Crunchie Flying Circus. Once the job was completed, he would return for another performance at Stourhead as the Red Baron on Friday evening and from there return to Dunkeswell. Saturday was the last display at Stourhead. Sunday would be a rest day. That was the plan. Whether Sheila would join him at Rendcomb for the Friday evening, or he return to Dunkeswell to work on the triplane was left undecided, and that was unusual. They had always worked out a plan of joining one another wherever he landed. The next day had always been planned for, even taking into account changes in schedule due to weather.

Thursday July 20th was also the day Sheila took her first flying lesson with the Devon School of Flying – a long awaited gift from Robin. She'd scheduled the lesson to allow herself time to assist him in manoeuvring the triplane onto the runway. Without brakes and tail wheel, the aeroplane – a near exact replica – was handicapped when it came to the act of simple taxi procedures. Sheila's job was to pull on the wings physically cajoling the machine into place before acting as guide for Robin who couldn't see beyond the three sets of steeply slanting wings when the aeroplane was on the ground; her role not dissimilar to that of the guide with the red flag in early motoring days. No wonder von Richthofen hadn't exactly taken to this particular kind of aeroplane. However, that said, airborne the triplane took on a different aspect. On the ground it was a goony bird always in danger of tripping over its own feet. In the air, it was an eagle.

Elated from her first lesson, Sheila found Robin finishing off a plate of sausage, egg and chips, and a cup of coffee in the airfield's clubroom. It was just before 7 p.m., enough time to do one more check. Then she remembered that they had still to make plans for the remainder of the week:

"If you get back to Rendcomb, ring me on your mobile. Don't worry if I'm not home because I'll be calling at Sarah's for a cup of tea."

Sarah, her eldest daughter, had recently married and bought a house in Ivybridge just two miles up the road from The Old Inn house that Sheila shared with Robin in Ermington.

"That's okay," he replied.

Robin stowed an overnight bag and tool kit into a fuselage compartment, poured 49 litres of Avgas into its single fuel tank and completed a routine pre-flight check. Always meticulous, he would often check his aircraft up to three times prior to every flight. Since buying it in 1984, he'd renovated the machine and later, after an accident in Germany, rebuilt it from scratch. He knew it inside out, literally.

And of the improvements he had made since the rebuild, he could now start the engine from inside the cockpit by electric ignition. Von Richtohofen would most definitely have approved.

"CLEAR!" he shouted. The twin bladed wooden propeller rotated stiffly at first as if awakening from a deep sleep – "CONTACT!" The tiny Lycoming engine spat into life starting first time as it always did. Strapped in the cockpit, Robin donned his leather flying helmet with goggles, adjusted the mixture until the engine revs settled with all pistons firing perfectly; it was time to leave. Guiding the tiny plane as it taxied toward the runway, Sheila watched his every move: the way he checked and double checked the controls and instruments; the way he sat high in the seat to see through the wings. Ready now, she pulled the triplane round by the port wing into what little wind there was. "I'll see you tomorrow!" she called above the noise of the engine. Standing, watching as the plane accelerated then quickly lifted into the air; standing mesmerised for what seemed like just a few moments though was probably longer, his old flying jacket – the one he'd given her – draped over her arm. Usually she would wave. She'd seen him off a hundred times and always waved him goodbye.

When all was said and done, this was her ritual. She always felt better waving even if he didn't always see her. Tonight for some peculiar, unknown, unfathomable reason, she didn't wave. Nor did she hear the aircraft waiting to take off behind her. Suddenly aware, she turned; saw its occupants looking at her as if to say: *Are you taking off too?*

Safely airborne, the triplane banked away east toward Stourhead; it was 7.25 pm. Peering over the cockpit, Robin could see the circular patterns of what were once parking bays for the Liberator bombers of the American forces, connected by taxiways and looking like some ancient design laid especially by some forgotten race for the gods to behold. To Lt. Jo Kennedy Junior and his fellow Liberator fliers, this aerial view would have been the most welcoming sight on earth. Flying back from anti-submarine and shipping sorties over the Atlantic and Bay of Biscay, in machines sometimes crippled, sometimes shot up, each man aboard praying to find his way home, across Lyme Bay desperately searching for familiar landmarks.

Climbing to just over 2,000 feet, Robin could see Lyme Bay to his right, some 12 miles to the south. He knew this immediate area well barely needing to consult the map. Flying at 80 knots, the flight was timed so that he would arrive as scheduled just before 8 p.m. His timing was crucial not only to rendezvous with the SE5s', but also to fit into the arranged schedule at Stourhead. The event was a Fête Champêtre to commemorate the centenary anniversary of the founding of the National Trust.

Stourhead gardens, the ancestral estate of the Hoare family and landscaped by Capability Brown in the eighteenth century boasted the finest mock Greco-Roman temples set amidst the most splendid landscape in the British Isles.

On each of the four consecutive evenings, the Fête Champêtre would stage significant happenings of the twentieth century in chronological order. The WW1 air battle entitled, *Death in the Evening*, was one of the first spectacles on the schedule, commemorating, of course, the years 1914 to 1918, man's disastrous penchant for war and the rapidly advancing technology of aviation.

Below him, the trees cast shadows across the green, undulating Devon pastures. Ahead, Somerset, with its larger, flatter, rectangular fields of ripening corn, barley and maize. Robin could follow the A30 trunk road for much of the way passing over the market towns of Chard and Crewkerne, then diligently obeying the Air Traffic Control over Yeovilton Royal Naval Air Station, its fifty mile radius defined in two shades of grey on the aviator's map. The Station, home to squadrons of Hawker Sea Harriers: jet aircraft so fast and powerful a collision could knock a reconstructed WW1 triplane out of the sky as if it were no more substantial than a box kite.

Beyond Yeovilton, Robin altered course heading Northeast over the Dorset county boundary for a direct path to Stourhead. At this point, he could see below him the fan vaulted roof of Sherborne abbey, its nearby monastic buildings now comprising a school, the evening sun illuminating their golden stone walls. From this height, the ancient, curving streets of the town looked like arteries flowing to the river Yeo. He could see people wandering around the ruins of the old castle where once Sir Walter Raleigh had lived. Beyond lay the farming villages of Poyntington, Milborne Wick, Abbas Combe, South Cheriton, Stoke Trister.

All along the route of his flightpath that evening, of those townsfolk, villagers and tourists enjoying the fine weather who stopped what they were doing for just a few, brief moments to gaze skyward would have seen one of the most unusual sights of the modern world. Of all the varied types of aircraft to pass through their busy South Western skies, this must have taken the biscuit. Seemingly, from out of a schoolboy's book of heroes; indeed, from the pages of aviation history, the Teutonic Red Knight of the skies was flying again.

In this neck of the woods, it was certainly not uncommon to see some strange machines flying in and out of Yeovilton. Concorde, the world's first and at that time only remaining supersonic airliner, had tested its engines in these skies in the late sixties. And that very same prototype 002 was now preserved in Yeovilton's aero museum. Fairy Firefly's, Swordfish, Buccaneers, Phantoms, Sea Vixens and Venoms had all flown this way at one time or another.

Looking to his left and some three miles west of Abbas Combe, he could see the ancient hill fort at South Cadbury – King Arthur's fabled Camelot. Below, the dual carriageways of the A303 mark the course of the old highway upon which coachmen had raced to break the record for horse drawn travel between Exeter and London. Galloping to stay ahead of the highwaymen too, though it was to be the railways that eventually caught them and finished their trade.

Ahead the densely wooded bluffs of what was once in Saxon times Selwood Forest. In what remains of the forest, and easily seen from the air, there stands at 160 feet atop Kingsettle Hill a magnificent triangular walled tower built in 1772 to commemorate the meeting of King Alfred the Great with his loyal supporters of Somerset, Wiltshire and Hampshire on Whit Sunday May 11th, 878 AD. From here, the King marched North-east leaving behind forever his island refuge in Somerset to defeat the Danes at Edington and reclaim England.

To the right of the tower, Robin's forward vision is impeded by the barrels of the triplane's twin replica machine guns. He needs to locate the church at Stourhead for his final bearing. And like an actor not wishing to be spotted before the curtain rises, he begins a circuit of the estate careful to remain out of sight of the crowds now filling the gardens below. Beyond, to the North-east, a line of cars queue patiently in the lanes leading to the fields allotted for parking – windscreens and sunroofs glinting in the sun. There is a light westerly air stream over the area. Ideal. Only two nights before, he'd marvelled at the distant rotund shapes of the Greco-Roman temples set around the shores of the glistening lake. Like his aeroplane, the temples are mock replicas of the mighty originals built by civilisations to the glory of the gods and success in war, the replicas below only a mere two centuries old yet they too commemorate a nation's glory in industry and war. How odd that there were such connections: replicas of classical buildings celebrating ancient empires; replicas afforded by the revenue of the new industrial age of the eighteenth century and Great Britain's own growing colonial empire. And above those buildings now, a mock aeroplane circles, symbol of another bygone age, its true predecessor a direct culmination of all that industrial might and technology, originally made not for vain glory or fun but wholesale destruction as the world's greatest industrial powers turned in on themselves.

Patiently Waiting

Forty-eight hours earlier, the first time he saw Stourhead, Robin knew he must return to visit at a later date with Sheila and couldn't wait to tell her so upon his return. At Dunkeswell, she had been waiting for him as she always did. He was fifteen minutes overdue. She assumed he was messing around, playing with his mistress the triplane as he was apt to do when the work was done. Eventually she spotted him coming in from the east. Nine o'clock, not quarter-to as he had promised. Other pilots were already putting their planes to bed; it was getting late after all. Mischievously, he lined the triplane up for landing, but instead of touching down put the smoke on, climbed and banked round to circle the field before flying through the smoke he'd just emitted encouraged from the ground by his fellow pilots egging him on, waving to him: "Go

Robin! Go!"

Safely down, he greeted Sheila with a big hug. They hadn't seen one another for ten days.

"We're going to have to go to Stourhead. It's beautiful! Beautiful from the air. We're going to have to go and have a walk around it."

A View from the Paddock

That second, soon-to-be fateful night of the Fête Champêtre, at the southern end of the ornamental gardens and well away from the crowds entering the main gate, Major Paul Tolfree, his family and friends, prepared an alfresco meal on a knoll in what is known as Turner's Paddock. From here they would get a splendid view of the flying and in relative isolation from the crowds. Only a few others had sited themselves in the paddock, among them Mr. Neil Andrew Richards and a few friends – members of the Shepton Mallet Young Farmers Club. Mr. Richards was later to recall at the inquest that Turner's Paddock offered "one of the best views."

Just before 8 pm, Robin turned the triplane south across the main lake. And, at which point, he made his last and only call over the RT. "Running in. One minute."

Chapter 2 - Doodlebug Summer

The enormous explosion demolished the buildings Violet Bowes had just been staring at through the upper window and shook the floor she stood on; its impact was close enough to rattle the window frames and might even have induced the birth, but thankfully didn't.

Had Lt. Jo Kennedy's mission to destroy the V2 launch site succeeded, expectant mother Violet would probably have remained at home in Sidcup. Kennedy's untimely death was now affecting the birth of another. After all, she and husband Harry had remained in their London home, 661a, Rochester Way, from the beginning of their marriage in 1940 and experienced it all from those first bombs in the early autumn of that year – Hitler's furious retaliation for the RAF's bombing of Berlin. From September 7th 1940 London had been bombed without respite every day usually with sirens sounding in the middle of the afternoon and again in the evening when the intensive night bombing would begin. This strategy of bombing London and not the airfields spared the RAF fighter squadrons and continued until November 2nd 1940.

From May 1941, conventional air raids resumed though they were more sporadic, but in the summer of 1944 the launching of V1 flying bombs meant that suburban targets across London and the Southeast could be hit with devastating results. Both the V1 pilotless flying bombs (colloquially known by Londoners as 'doodlebugs') and later the more advanced V2 rockets could deliver an extremely destructive warhead capable of taking out entire neighbourhoods. The V1's destructive power was 'enhanced' unintentionally by a design flaw in the jet motor that powered it. The pilotless plane was designed to power into a dive that would have created a crater on impact. However, a fuel starvation problem caused the motor to cut whereupon the bomb would glide to the ground and explode as it hit the surface thus causing an even bigger impact than had it pushed into the ground under power.

Those Londoners that survived would remember 1944 as 'The Doodlebug Summer'. The Nazis were aiming to launch 200 attacks an hour on the capital; that on occasion they could achieve 200 a day was in itself a devastating setback to both the civilian population and armed forces as they embarked on the final chapter of war.

Large urban areas of south London were being devastated with indiscriminate attacks finding their targets. For Harry and Violet Bowes and their baby daughter Christine, the risk was becoming too great with boroughs such as Greenwich and Woolwich receiving some of the heaviest strikes. For the duration of four, long years of war the couple had witnessed the worst the Luftwaffe could deliver, but the V1 attacks were bringing a new terror to the suburbs. Public buildings, factories, churches and housing were either destroyed completely or rendered uninhabitable through bomb damage. Even lesser damage by way of broken windows and roof slates in those buildings furthest away from the epicentre of the blast (usually beyond a 600 yard radius) could mean misery for the inhabitants in that unusually cold, wet summer when materials for repair were hard to come by.

By July, 15,000 Londoners a day were evacuating their homes – most on packed trains as even the most stoic and war hardened of people came to the collective realisation that to remain would be foolhardy. Under these extreme conditions

Violet reconsidered Harry's advice: if possible, get a ticket and take the next available train north to Blackpool and stay there for as long as it takes.

Robin Austin Bowes was born on the 25th August 1944 in the relative security of the Kimberley Hotel, 8, St. Chad's Road, Blackpool just as the allies manage to overrun and destroy the sites used by the Nazis to launch the V1 attacks. On September 7th Duncan Sandys announces that the Battle of London is over, though his statement is outrageously premature as the first V2 rocket hits Staveley Road, Chiswick the following day.

The Dawning of a New Age

Almost 9,000 people lost their lives during that Doodlebug Summer with thousands more injured while others are affected by the loss of loved ones, family homes and businesses. Sometimes as many as ten doodlebugs would arrive together over urban areas despite valiant efforts by both anti-aircraft gun crews on the ground and RAF fighter pilots to destroy these vicious weapons in the air. The V2 rocket attacks that began in February of 1945 continued the onslaught that, despite allied advances on the continent, was demoralising Londoners and steadily draining the capital of its populace.

Post-war rhythms

By 1950, the people of Blackheath were gathering again at the Prince of Wales Pond, largely to admire the exquisitely crafted model sailboats. For Robin and his elder sister Christine, a trip to the pond with their father, Harry, was a treat, although adult modellers dedicated to their hobby did not always appreciate the launching of less sophisticated 'toys' by small children.

Young Robin Bowes, however, was not about to make any

waves on the pond. Robin was already a dedicated four-wheel enthusiast and rarely ventured anywhere without a toy car in his hands.

Robin in pedal car

His pedal car – made with love by Harry – was the most beloved possession of all and at an early age he was soon impressing everyone with his ability to carefully manoeuvre the car using three-point turn techniques and generally taking great care not to scratch or dent its sturdy, metal body.

His ability to improvise was no less impressive, but on one occasion it came close to bringing his new found driving abilities to an untimely and tragic end. Positioning his pedal car behind the closed doors of a van delivering the week's groceries to the family, Robin – with all the naïve confidence of a four-year old – tied the front of his car to the back bumper of the van; then sat back and waited for take-off along the Rochester Way. It promised to be the ride of a lifetime and had it not been for mum Violet glancing out of an upstairs window at a critical moment, the grocer – having completed his delivery – would have been oblivious of the situation directly behind him and thought nothing of pulling away and out into the traffic unaware of the young adventurer in tow behind him.

Line drawing of Bentley Blower by Robin Bowes

Robin's inspiration for all things motoring came from his engineer father Harry who was a keen driver and car enthusiast who enjoyed taking his young son to post war race meetings at nearby Brands Hatch to watch the popular J.A.P. engined Cooper racing cars. A keen amateur cine camera enthusiast, Harry would capture on 8mm film the rising stars of the fifties such as Mike Hawthorn and Stirling Moss – eager young sportsmen making their names in the Formula III 500cc class.

Both sides of the family (paternal and maternal) were musical in talent and spirit and so those drab, dark post-war years of shortages and hardship were brightened by the love of playing and performing whether it be Harry and his brothers Bill, Albert, George and brother-in-law Bob, performing at a local venue or at Christmas family gatherings when grandparents, aunts, uncles and cousins would all come together for seasonal parties of games and music around the fireside.

Not surprisingly in this pre-television atmosphere of genuine home entertainment, Robin's ear for rhythm and beat marked him out as the latest member of the family to have a strong musical bent – particularly in respect of timpani.

Sundays were for dressing up in best clothes and promenading. With the warm days of spring and summer, Harry and Violet would take the children to walk in Greenwich Park and marvel at the Observatory and the magnificent Cutty Sark laying redundant and dormant in its dry dock. Wondering of its epic sea journeys the family would soon be embarking on their own sea journey that would take them away from their native Greenwich to a Mediterranean island.

Malta

Line drawing of Valetta Harbour, Malta by Robin Bowes

London was already beginning to change in those post-war years. The close-knit 'extended family' communities of parents, aunts, uncles, cousins and work colleagues that had been in place since the dawn of the Industrial Revolution were now fragmenting as job opportunities in different parts of the country and the world at large opened for those with skills, experience and at the very least, a willingness to work for an improved life.

In place of the terraced housing new tower blocks were emerging on bomb sites and craters left by five long years of bombing. New towns such as Harlow, Basildon and Stevenage were under construction replacing the once green fields of the Essex countryside. Londoners, like so many other British, who had for so long known no other environment were adapting to the idea of new horizons and distant opportunities.

For the Bowes family, Malta was one such opportunity. Harry's engineering abilities were needed in the naval dockyards at Valetta the island's capital. It would be a two-year posting and a perfect opportunity to enjoy the bright blue skies of Mediterranean life in contrast to London's depressive winter gloom and smog.

Arriving in Grand Harbour, which then, as now, was busy with boat traffic of all shapes and sizes, was a welcome sight after nearly a week on board the troopship Empire Windrush. Robin, then five, and Christine, eight, had enjoyed themselves immensely with the various fun activities put on especially for the younger passengers by the crew, but it was still wonderful to see for the first time the Mediterranean island they would call home for the next two years.

Harry bought his first 8mm cine camera in Malta to record family life in their new home and around the island – on swimming trips to the many golden beaches and the various cultural events. The beaches provided a perfect opportunity to teach the children to swim, an ability Robin eventually picked up though his preference was always dinky cars.

Like London, Malta was changing in these post-war years. Although the tiny island had long been a close friend and staunch ally of Britain providing a base from which the Royal and merchant fleets could operate, by 1950 the Maltese were embracing British culture and language like never before. English was being taught widely and progressively in the schools; Maltese parents were using English when talking to their children; street names and shop signs were anglicised as English words were incorporated into the Maltese vocabulary. English language books and newspapers also came with the new wave of British coming to help rejuvenate the island following the bitter years of siege.

Field Marshall Erwin Rommel had so badly needed the island to consolidate his hold on North Africa, but the valiant

Maltese with the aid of the RAF had seen him off, though at a terrible cost. Had Malta fallen to the Axis powers, General Montgomery's eventual victory at El Alamein might never have come.

Malta's new Labour government elected in 1947 by proportional representation were keen to reconstruct the island and in so doing boosting employment and improving social services. Britain's RN and RAF presence was welcome but the island's leaders were concerned that as the economic state of Britain had been left so perilous after five years of war, that economic restraints on defence spending could mean the withdrawal of defence for the island at anytime particularly as military strategists in line with NATO thinking were forecasting future naval conflicts with the USSR occurring in the North Atlantic rather than the Med.

By the time the Bowes family arrived in 1950, the Socialists had won back power and some observers feared a rather more pro-Italian stance, but Prime Minister Dr. Enrico Mizzi's term was short lived. He died in office only three months after the election and was replaced by Dr. George Borg Olivier, though without a working majority at that time.

The Bowes family were not the only London family entering Malta during that period; another family with a strong London connection, and ironically a Bowes connection, too, had gathered on the island. Newly married Princess Elizabeth rendezvoused here with Prince Philip who was commanding His Majesty's frigate Chequers. His uncle and aunt Lord and Lady Mountbatten had returned to Malta after an absence of ten years and played host to the couple at their newly refurbished home at Villa Guardamangia, barely a short walk from Harry and Violet's flat in Pieta.

Malta Part II

Line drawing of Gloster Gladiators over Malta by Robin Bowes

The family's second posting to Malta came in 1958. The Island's government under Dom Mintoff was already making plans for independence from Britain, but for the next couple of years at least, Malta's defence remained reliant on Whitehall.

For the family it was like returning to see old friends and familiar places. Robin, now 14, attended the Royal Naval School at Tal Handaq and mixed with the children of service families stationed on the island. His familiarity with the island meant that he was much more able to strike out on his own cycling and teaming up with mates. Tal Handaq was a good school allowing plenty of opportunity for extra curricular activities in music and sports where he did well; and at weekends he'd assist Harry in the making of cine film adventures where they'd painstakingly piece together a narrative for home entertainment.

Chapter 3 - A New Life in Devon (1961-1967)

Line drawing of Hawker Hunter by Robin Bowes

During those years Tom Wolfe wrote in the introduction to his first book, The Kandy-Kolored Tangerine-Flake Streamline Baby: 'Stock car racing, custom cars – and, for that matter, the jerk, the monkey, rock music – still seem beneath serious consideration, still the preserve of ratty people with ratty hair and dermatitis and corroded thoracic boxes and so forth. Yet all these rancid people are creating new styles all the time and changing the life of the whole country in ways that nobody even seems to bother to record, much less analyze.'

You would have to agree with Wolfe back then, that at the beginning of the decade, no one knew what to call 'it'. Kids racing stock cars around dirt tracks, strumming rifs on electric guitars, dancing the 'jerk', the 'monkey' and the 'shake'. The fifties was gone – bye, baby bye! Nice knowing ya! Stay cool daddy-o! Far out man!

British New Wave cinema picked up the baton from French Nouvelle Vague; Don McCullin snapped pictures of dead soldiers and starving children for the new colour supplements;

the Beatles returned from Hamburg and Robin Bowes bought a stock car – painting it yellow and blue then driving it in ferocious circles against thirty other old jalopies in a scene that looked less like a sport and more like a movie director's recreation of Bonnie and Clyde's last chase across Missouri.

When Wolfe wrote of some American place, he could have been writing of some English place, too. Somewhere like Plymouth. At the beginning of the decade it was as if a tidal wave had begun to build on the eastern seaboard of the United States; a wave sweeping east, building power, gathering momentum with culture – lifestyle, fashion, diversity – surfing its crest. Ironic that what England had exported some 300 years earlier was now returning: the ingenuity that had been the bedrock of 17th century Europe was crashing right back into Plymouth's shores!

Concrete shopping plazas, department stores, supermarkets, a suspension bridge like the one in San Francisco (only smaller), big cars with chrome, rock and roll music and stock car racing. One wonders whether the Pilgrims with their dreams of a new world ever realised just how much they would change their native land and the port from which they embarked.

It is worth mentioning Wolfe, not because this was a time in which he was making his mark as a writer, but because he witnessed and noted something that we might otherwise have forgotten or be unaware of all these years on. That is, that sports like stock car racing in America's south (and England's south) were so new, they were marking the most remarkable bloodless revolution in the history of Western culture and the planet in general. No poor kids had been rich like this before. This generation was upwardly mobile, but these were ordinary kids from the streets – kids like Robin Bowes.

Robin with chequered flag

His apprenticeship at Hawker Siddeley abandoned, he had joined his parents in Plymouth where Harry was now posted. The West Country was almost as much a different country for the family as Malta had been. Although just 200 miles away from the south-east suburbs of London, Devon culture was very different and, like Malta, it offered a wealth of experiences with its beaches, coastline and nearby moorland.

Despite the distance, Robin was keen to keep in close touch with friendships that he'd made while at school in Malta.

By September 1962, Robin's fellow Talhandaq classmate, Roger Wilkin, had also returned to England with a view to making the RAF his career, but until he was old enough an office bound job with Commercial Union Insurance in Southgate, London, would have to suffice. At least it was a job where he could earn a wage whilst kicking his heels until reaching the required minimum age of seventeen years, six months – the minimum required age for pilot training. In his spare time and with whatever cash he could muster he'd head for Paddington station and board a train bound for Plymouth his return ticket in one hand and his treasured guitar in the other. Roger and Robin had not only been classmates; they'd been bandmates cutting their teeth playing local gigs around the island. Now, in an effort to re-capture the heady days of Malta the boys were always keen to join together to play rock 'n' roll music whenever the opportunity presented itself.

In the Easter of 1963 a Tal Handaq School Reunion was arranged at Eastbourne: the perfect opportunity to load Robin's drum kit and Roger's guitar into a mate's 1935 Morris 8 – the 28-year old veteran motor being expected to cover the 200 miles to Kent. Every 30 miles or so Robin would have to coax the fuel pump back into life as they navigated their way through the many towns and villages of southern England in an age before motorways and by-passes.

That same Easter, Roger was accepted by the officers and aircrew selection centre for pilot training and having resigned his post with Commercial Union he spent the last month before call-up accommodated as a very welcome guest with the Bowes family in Plympton. Robin's own interest in aviation was blossoming during this time. Having abandoned his shop floor apprenticeship with Hawker Siddeley in London he was now employed as a car salesman with A.C. Turner at Derry's Cross and with his earnings was learning to fly at Plymouth's Roborough airfield under the watchful eye of his instructor, Mr.Lucas. While Robin was learning to fly in his spare time, Roger was undergoing the full works of regimented training as an RAF Officer Cadet initially at South Cerney in Gloucestershire and following graduation as a pilot officer training on jet provosts at Leeming in North Yorkshire. Each month the rookies were allowed a long weekend break giving Roger the opportunity to catch the train to Plymouth where Robin – dependable to the last – was always there to meet him.

As a junior car salesman Robin would often have a car for the weekend for the purposes of demonstrating to a prospective buyer, and if Roger was in town then they at least had transport for a trip around town or up to Dartmoor – providing they could pool together enough cash for petrol. If Harry's car were available Robin would demonstrate to Roger his new ability in executing handbrake turns in the narrow terraced streets of Pennycomequick. Not that Harry was ever aware of the uses to which his car was being put as it was always returned in immaculate condition.

Sat on bonnet of Aston Martin DB2/4

A new breed of sportsman

Driving an ancient Morris 8 gutted of all its interior fittings save driver's seat, steering wheel and dashboard, Robin began to make a name for himself as a stock car racing driver.

The Morris 8, its days as a workhorse over, was to finish its days in Pennycross Stadium – the motor sport equivalent of a gladiatorial ring. Stripped of all former glory its doors welded shut, the once upright and compact body was reinforced to roll in the dirt. There was to be no more cruising the Queen's highway for this compact marque of pre-war British motoring.

Climbing out of a Morris 8 stock car

Fearnley Ough, a Saltash garage owner, maintained the car, but this arrangement was costing Robin dear as Saltash was separated from neighbouring Plymouth by the expanse of the river Tamar. Prior to 1961 and the building of Plymouth's first suspension bridge, Saltash and the rest of Cornwall were only accessible by rail and ferry services. The new Tamar bridge ran alongside Isambard Kingdom Brunel's superbly engineered railway bridge, which for over 100 years had been the only bridge crossing the mouth of the river.

Of course, the completion of the suspension bridge meant there was a price to pay for its investment. There was no cost for entering Saltash and Cornwall, but there was a tariff for crossing back into Plymouth and it was these return journeys that were costing him so much. Mounting the car on a trailer

Robin would tow it across the bridge for Fearnley to paint the roof either blue or yellow depending on how Robin had qualified the previous race.

Not one to complain, it was a price he was willing to pay, but he soon saw a way around the issue. Instead of thrashing the jalopy he would invest in a larger car with chrome aplenty – a luxurious Vauxhall Cresta no less with a two-litre engine. In its class, it would win hands down earning Robin accolades as not only a winning driver but for having the guts to risk entering such a prestige car in what had been the preserve of modified wrecks.

Not everyone was so enamoured by his new acquisition. One night, a group of disgruntled track jockeys menacingly surrounded his car at the end of another successful race. He had the good sense to sit tight and push down the door locks while his tormentors rocked the car and kicked its tyres. Later, Roger questioned his friend's motives:

"You've got this wonderful car, it's a super saloon car, it's your prime means of transport. What on earth are you doing racing it as a stock car?"

"This way I can be sure of winning!" Robin replied.

And there were other ways of winning. Away from the dirt track there was far more to cars than the frenzied pedal-to-the-metal dash and crash tactics of stock car racing. The subtleties of driving skill competitions also appealed. Precision manoeuvring at slow speed whilst engaging the smoothest of gear changes over a set course that incorporated obstacles such as bollards appealed every bit as much, and soon Robin was adding this ability to his growing credentials. He was winning plaudits as not only an extremely fast driver, but a thoroughly proficient slow one, too.

A fellow car and music enthusiast, Colin Goss, introduced Robin to the delights of Aston Martins. In the 1960s, as now, there was no greater pose value than an Aston Martin especially as the type had been made famous in the James Bond film, Goldfinger. On numerous occasions Robin was able to borrow Colin's DB2-4 in return for reciprocal favours. Some years later he purchased his own DB2 – a near immaculate car save for one small flaw: a hole had been cut into the leather panel of the passenger door for the purpose of fitting a radio speaker. For some reason, the idea had been abandoned and so a patch covered the offending hole, but it was always noticeable and irked Robin all the days that he owned the car.

Chapter 4 - By the Way, I've bought an aeroplane ...

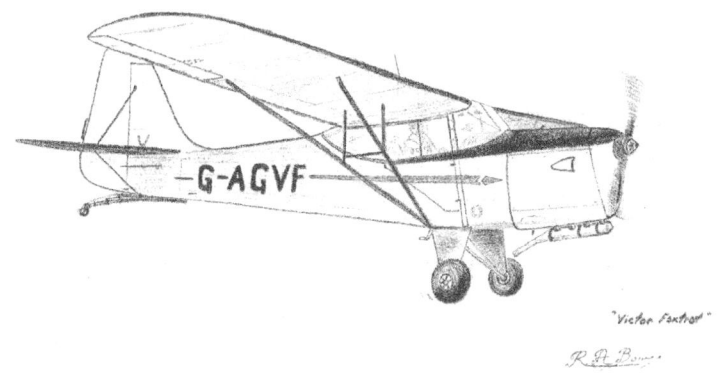

Line drawing of Robin's Auster by Robin Bowes

Having gained his Private Pilots License under the expert tutelage of Mr. Lucas at Roborough, Robin looked to buy his own aeroplane. He'd had his eye on an old, ex-military Auster and with his new PPL he was in a position to make a purchase.

Drab in appearance with a simple cockpit layout, the austere Auster had been made purely for military use. Developed in a pre-helicopter age, the type was ideal for operating from short grass strips close to the front line where its role as a spotter plane enabled a pilot or observer to monitor enemy troop movements or alternatively used as an airborne cab ferrying personnel. With its V-strut wing bracing and parallel chord wings, the workmanlike Auster was every bit as unpretentious as the Morris 8 stock car had been. The white, multi-striped US built Piper Cubs and Cherokees that chiefly comprised the Roborough flying club were extremely expensive and therefore often the preserve of recreational flyers – entrepreneurs who had built their businesses in the post war years and were now enjoying in their middle years a more prosperous Britain. This was a world where acquisition was everything, where the consumer was king, where the product could be as humble as a 45-inch vinyl single bought in the High Street or as prestigious as a bespoke four-berth cabin cruiser or four-seat single-engine aeroplane built not for war, but for personal pleasure. Post war austerity was over and in an age of glossy colour the little Auster was always going to be overlooked by the more affluent club members; but for a young car salesman the Auster was the only thing in reach with values ranging from £500 at the bottom end of the market making the type the only serviceable aeroplane Robin could afford to buy. And it wasn't simply a matter of cost: the history of the aeroplane appealed immensely; it was a working pilot's machine made for war not pleasure.

Golf – Alpha Golf Victor Foxtrot was also an ideal introduction to the world of historic aeroplanes. Although essentially modern and unprepossessing in design, the Auster was, and remains to this day, no piece of sweet cake for the novice pilot. It's an awkward little machine that presents its handler with attitude from the outset.

Originally manufactured by the American Taylorcraft Corporation before being licensed for production at Leicester, the Auster continued to be manufactured after the war as a light aeroplane for the civilian market, so not all the type saw war service.

The cramped environment that is the cockpit with its quaint Bakelite switches gives the control surfaces a rather art deco feel. Two long, helicopter like control column sticks rise up from the bell crank arrangement. The rudder pedals are situated above additional heel-operated brakes that make it very difficult to operate independently of the rudder requiring a dexterity of the feet when taxiing.

In flight, attaining a decent trim requires adjusting a lever situated in the roof panel and demands frequent attention if the nose is to be kept on the horizon. The throttle lever and mixture control are easily manipulated by the right-handed pilot and there's excellent visibility thanks to the over wing configuration – a feature that made the type so proficient in the role of "spotter" aircraft.

Depending on its power unit, the Auster can be slow to take off and once airborne its control surfaces are heavy and begrudgingly unresponsive. The pilot has to cajole and put pressure on the controls to initiate turns. Furthermore, the Auster lands like a rubber ball bouncing along the grass. Approaches need to be low and incredibly slow, but at least it's forgiving in its stall speed at just 38 mph and this is what makes it such an effective STOL aeroplane.

In fact, the Auster was an ideal aeroplane for Robin's future career as it acted as a flying classroom helping him to hone skills that he wouldn't have learned at the controls of an easier, more modern type.

When Roger looked through the viewfinder of his camera and snapped Robin proudly posing in front of his new acquisition gazing skywards, donning his flying gloves, he could have had little idea of the career path that his friend was embarking upon.

News of Robin's acquisition of both PPL and Auster came as a total surprise to Harry and Violet; in fact, it was something of a shock at first, especially as they weren't even in the country. Harry had been assigned to the giant Lockheed Company in California for a three-year contract and they were living in the district of Sunnyvale some forty miles south of San Francisco.

First acquisition: Auster: Golf-Alpha Golf Victor Foxtrot

One evening around teatime, Harry answered the phone: it was Robin calling as he often did to bring his parents up to date with his news at home. Almost as a matter of course, and sometime after the customary preliminaries of 'How's mum? … Good! How are you dad?' He just happened to mention: 'Oh, by the way, I've bought an aeroplane.'

This announcement in itself wasn't so unusual; they'd remembered his love of making model aeroplanes as a boy, but they hadn't considered that he would pursue a greater ambition to actually fly himself and there was no history of aviators in the family with not one aviator among them; at least, not until now. So Harry's immediate reaction was simply: 'Oh, is it a model?' Thinking it to be a radio controlled model he'd just purchased from a model shop in the city centre. 'Oh no,' Robin assured him, 'it's a real one. It flies.'

'Well, what about you?' asked Harry.

'Oh yes, I've got a pilot's licence.'

To Harry and Violet, who, up unto that moment had known nothing of their only son's aspirations or pilot training, it was a revelation indeed. When asked how he could have afforded it all on a car salesman's wages he informed them that he'd won a special scholarship for the most promising pupil, but the bulk of the costs were at his own expense.

It wasn't long before Harry was able to see for himself during a fortnight's break at home in England. Robin was not only keen to show his dad what he'd bought he was also keen to take him up for a flight; and Harry, whatever his initial reservations, cheerfully agreed to a flight over Plymouth.

When Harry and Violet returned to England a few years later, Harry's new posting with the Admiralty in Bath meant that they would be able to see at first hand Robin's achievements as a pilot. On a trip to visit his parents in Bath, Robin asked Harry if he thought mum would agree to flying with him in the Auster back to Plymouth. In principle, this was okay with Violet, but in reality the weather was not set fair at all; and for someone who had never flown in anything smaller than an airliner, Violet, strapped tightly into the observer's seat, could be forgiven for having second thoughts as this tiny, noisy little aeroplane waited for the winds to subside enough for take-off. When they eventually eased and the aeroplane climbed into the wintry skies above RAF Colerne Violet could not help but be impressed by Robin's skill in handling the little machine so expertly and in such difficult conditions. As they headed Southwest across Bath and North Somerset her initial anxiety eased as he pointed out various landmarks, but the further west they headed the worse the weather became as they encountered first rain and then snow over Dartmoor. Caught in the middle of a snow storm unable to go around or above, Robin reassured his mother that all was clear at Roborough just twenty miles away. Violet turned to him and smiled sweetly – she had every confidence in her son's newfound abilities.

Infamous districts, sleazy joints

In order to fly and maintain the Auster, Robin would work all the hours he could to buy fuel and to cover the cost of insurance and house it in a hangar. He was finding to his cost that flying was by no means a cheap affair; his full time job of selling cars by day needed to be supplemented by playing the drums in a band at night often to midnight and beyond. At least he had a talent for music that he could fall back on when he needed extra cash and he loved drumming so there was never a need to drive a taxi-cab or work behind a bar, though some of the inner city venues he found himself playing in lacked the ambience he might otherwise have chosen for a night out. And for all his hard work when there was enough cash for a quick flight there was the rigmarole of pre-flight prepping – including the manhandling of other machines to clear a space to get the Auster out of the temporary canvas hangar where it would invariably be trapped at the back, then pushing them back in again once the Auster was clear. It was always easier doing this with a companion, especially when that companion was his old chum Roger Wilkin – himself a newly qualified RAF pilot stationed at West Rainham.

As a drummer, Robin earned his money playing in some of Plymouth's more nefarious and seedy joints in and around Union Street. Some gigs were in real dives, but the band never let a venue get in the way of the music. There was never a temptation to rush a tempo or cut a chord. Robin's drumming was noticeable for its delicate touch and both band and audience alike appreciated it. Despite the often confrontational atmosphere of rowdy behaviour and raucous noise going on through the playing he was never tempted to simply compete with the noise and bash seven bells out of the drum skins. His final touch was often to incorporate the ring pull at the end of the brushes running it across the crash symbol in the final beat of the number *Szzinngg*! causing the symbol to reverberate – ringing and ringing and ringing. Not once was he ever tempted to smash and bash his beloved drum kit for the sake of attention: *Look at me! Loud is great!* Precision was his watchword and it remained so in everything he did.

Robin on drums

On occasion, if the band were a musician short, such as a rhythm guitarist, Robin would recommend Roger as a stand-in and although initially he would have problems keeping pace with riffs and chord changes, once he'd settled in the experience could be enjoyed. Robin appreciated the company of a mate and a contemporary in a band comprising largely middle-aged musicians.

Robin's dedication to flying the Auster meant that spare time and cash for socialising was at a premium, but he would still find time for girlfriends and old mates like Roger. Always keen to try new things one girlfriend even introduced him to horse riding, which he took to. In fact, the young Plymouth bachelor was always a popular companion, but for the time being he was not prepared to commit himself to a serious relationship, despite his tenancy with a landlady in Whittington Street, despite the house having no bathroom or inside toilet – ablutions required a tin bath. Robin was in no hurry to settle down.

Bacarrach & David days

Wendy Harley, a 24 year-old local girl, worked as a loggist with Westward TV the ITV regional broadcaster for the Southwest of England. The studios were right in the heart of Plymouth at Derry's Cross just across the road from A.C. Turner's garage where Robin worked as a car salesman. Wendy's role was to log the timings of programmes and commercial breaks – in fact anything and everything that happened between the times the station began broadcasting each morning to close-down around midnight. Incidents might include any breakdowns in transmission, for example. In a medium of split second timings, Wendy's job was crucial in ensuring the smooth running of daily broadcasts that incorporated live and recorded programming.

One evening over drinks, a family friend and colleague of Robin's, was so effusive in his praise of Robin's personality and talents that he suggested that Wendy should meet him; and so a date was arranged.

A dance at the Victoria and Albert at Stoke Gabriel provided the occasion. Travelling in the back seat of Robin's Aston, Wendy could see little of her suitor – only his eyes in the rear view mirror. His quiet demeanour, coupled with an old-fashioned, courteous attentiveness attracted her immensely. Upon arrival, stepping out of the car she could see him to be a most handsome, solidly built young man, immaculately dressed and so very attentive. Robin Bowes more than lived up to his reputation.

Two weeks later, she was on a sleeper train heading east for London with Robin in a neighbouring compartment and the Aston safely secured on a car freight wagon at the rear of the train. For Wendy this was a rare excursion out of Plymouth let alone outside of the West Country and all for a man she barely knew and for a type of event she had little idea existed.

Long before they'd met, Robin had posted his entry form for a Concours d'elegance competition and time trials as part of an Aston Martin Owners Club weekend to be held in the Midlands. The weekend was a perfect opportunity for he and Wendy to become acquainted and if conversation became a bit strained they would at least be meeting up with his sister Christine and her husband, Ralph at Paddington station where the car would be unloaded for the final leg of the journey – the drive north up the M1 to Birmingham.

Christine and Ralph also had no idea just what a Concours d'elegance competition was, and neither had they seen the glamorous Aston, so resplendent in its dark blue livery. According to Club rules the one hundred mile journey up Britain's premier motorway was a necessary requirement of the Concours event, in that for the class it was entered into the rules stated that it had to have travelled at least a hundred miles under its own power. This was a dream of a ride in such a prestigious car – a type that was, thanks to the Bond films, already becoming something of an icon in sixties Britain. The films had marketed Aston Martin far more successfully than the tiny company could ever have hoped to do with an advertising agency. Indeed, so iconic had it become that demand was outstripping supply with a lengthy waiting list for the Newport Pagnell works where every car was hand built by the most experienced and capable of Britain's engineers and craftsmen.

As powerful and as glamorous as it was, the Aston was not designed to accommodate four adults travelling a hundred miles. There was no alternative but for Ralph to travel in the front passenger seat whilst Christine and Wendy squeezed into the superbly upholstered but cramped rear seats. There was much catching up to do in respect of family matters and getting to know Wendy, but in any case, it wasn't all family talk. There was a dinner and dance to look forward to on the Saturday evening giving them an opportunity to dress up in their finest attire. The drive to the Midlands wasn't entirely without incident: the sight of smoke whisping out from under the bonnet caused a certain amount of consternation, but Robin remained remarkably composed. Lifting the long, heavy bonnet, it seemed that the type of paint he'd used was not suitable for the high temperatures of the engine compartment. At least the incident gave Wendy, Christine and Ralph the first opportunity to stretch their legs and marvel at what was the most impressive car engine they'd ever seen. Not only was the motor extremely large requiring its feed of premium petrol from no less than three SU carburettors – it positively gleamed in the sunlight. They could practically see their faces reflected in the mirror-like chrome finish of the mighty twin rocker box covers. Indeed, the whole compartment was spotless – totally devoid of any traces of grease or dirt normally found under the bonnet of an average car. Of course, this was no average car.

Wendy was quick to notice not only the amount of effort Robin had put into the engine and indeed the appearance of the whole car, but also the manner in which he dealt with what might have been considered a crisis. Just prior to the beginning of the event as he was touching up the word Dunlop on a tyre wall some of the yellow paint spilled onto his smart suede shoes, but he remained calm and collected – not one profanity passed his lips.

Although precious about the Aston's appearance, for Robin, this was a car for friends and family to enjoy and sharing the experience gave him great satisfaction.

There were various classes making up the Concours and Robin was looking to score maximum points for the car's overall cleanliness, condition and originality. Judges could award up to 100 points for the best engine compartment, chassis, body exterior and body interior; but in Robin's case, the patch in the door lining made it unlikely and so he didn't hold his breath.

There was greater success to be had the following day in the sprint. In those days, the DB2/4 MKII had an impressive 0-60 mph capability of just over 10 seconds, and Robin brought to bear his experience of grid starts honed on the stock car circuit and hill climbs.

By the time Robin and Wendy were en route to Devon they agreed it had been an unforgettable weekend – especially for Wendy. He had been the perfect gent in every respect especially considering he'd inadvertently seen her in curlers and without make-up on the sleeper train; and even when they stopped at services on the A38, her faux pas of placing a plastic cup of lemonade on the roof of the Aston accidentally spilling some of the contents into the car's immaculate interior, was not enough to upset Robin's gentle patience – he put her at ease immediately. She could now most definitely see that Robin Bowes reputation was by no means exaggerated. In the coming months she would see the numerous other sides that made up the many faceted character of the man she would fall in love with and marry.

Flying by the seat of their pants

Saltash businessman Pat Crawford had his very first flying lesson in the Auster that Robin eventually bought with the proceeds from the sale of the Aston Martin. A former RAF national serviceman Pat had funded flying lessons through his business as a motorcycle dealer qualifying for his PPL at Roborough shortly after Robin had gained his. Shared interests in cars and a love of flying brought the two men together and they were soon owners of aeroplanes – Robin with the Auster and Pat with a French built aeroplane. They would often fly their machines together or they'd team up in Pat's machine for a jaunt down the coast or up country. Never satisfied with one particular type, Pat frequently changed his aeroplanes thus enabling Robin the opportunity to fly each new acquisition with the result that he soon became experienced with various types.

One type that Pat particularly aspired to was the Belgian built Stampe SV.4 – a biplane of the 1930s' and in much the same mould as the British built de Havilland Tiger Moth so loved by their former instructor, Bill Lucas. Like the Tiger, the type was well known for its excellent acrobatic qualities and in the late sixties was outclassed only by more modern rivals such as the American built Pitts.

When Pat eventually acquired a very fine example of the type, he and Robin would fly together but only after a toss of a coin to see who would sit in the front and who would sit in the back of the tandem seat biplane. With no canopy or radio they would use gesticulations to communicate just as the early aviators had done. However, this lack of modern sophistication led to one near disastrous occasion. Pat had lost the coin toss and as a result sat in the front; their flight plan was simple enough – do a few circuits, about two each, whereupon whoever was pilot would "shake" the stick to indicate to the other the handing over of control. Thick cloud was preventing them climbing above the circuit height, but when a break in the cloud appeared Robin was quick to take advantage climbing to a height where the machine could be put through its acrobatic paces.

Dipping the nose he put the Stampe into a dive before pulling it up into a loop totally unaware that Pat wasn't wearing his four-point harness. Gripping the interior edges of the cockpit for dear life as Robin came out of the loop, Pat relaxed his grip enough to turn round and gesticulate and mouth the fact that he wasn't belted in. 'Yeah, okay,' mouthed

Robin in reply and promptly dipped the nose to put the aeroplane into a second loop. It was only the law of physics that kept Pat within the cockpit. Pushed in by centrifugal forces it was fortunate that Robin hadn't rolled the aeroplane.

When they eventually touched down Robin now sensing that all had not gone well was able to enquire of his co-pilot: 'What's up?'

Pat shouted at him, 'You bloody fool, I was trying to tell you I didn't have a seatbelt!'

It could have been the end of a brief but happy friendship.

Making Films

Film maker and musician Gordon Clarke first met Robin when he bought a large Ford estate from Evans and Cutler. Robin was one of those quiet, understated salespeople who could sell simply by helping the customer find what it was they were looking for with no pressure; and Gordon, like so many people who met Robin was won over by his kind and helpful personality.

Through conversation, Gordon and Robin discovered a shared talent in that they were both drummers playing with local Plymouth bands. Gordon was also an amateur film maker and he asked Robin if he would do some driving for a film he was making entitled *Dominuendo*. Robin was delighted and also suggested that Gordon might like to consider making use of his Auster. The inclusion of an aeroplane could really make a film dramatic!

Come the day of the shoot, Robin brought along a beige coloured Rover 2000 P6, which was ideal for Gordon's requirement.

'We're going to shoot this scene at tarmac level,' he told Robin.

'What do you want me to do, Gordon?'

'Drive as fast as possible around a bend towards me.'

Gordon didn't tell him exactly where he'd be with the camera only that he could pass him as close as he dare. Robin, picking up speed, entered the bend unaware that out of sight on the other side of the corner Gordon was lying prostrate and exposed on the tarmac directly in the path of the speeding Rover. Holding his nerve, trusting Robin's abilities as a driver Gordon was able to pull off a shot that was to become one of the most dramatic sequences in the film.

Relieved that he hadn't left tyre prints all over the director's face and back, Robin parked the Rover, walked over, smiled and enquired of his new friend:

'Was that close enough?'

Following the success of the ground-to-car shot, Gordon considered Robin's suggestion of bringing in the Auster for the film's climax.

Gordon's direction required his central character played by a young actress to run toward a cliff edge carrying a case loaded with stolen money. Running for the cliff she is harangued by the Auster as it swoops low above her head causing her to fall and drop the case. The scene is not unlike the famous crop dusting scene in Hitchcock's *North By Northwest* where Cary Grant's character is chased into a maze field by a Boeing Stearman biplane.

With each take, Robin at the controls of the Auster would come ever lower in his attempt to drive the girl off the edge of the cliff. As the actress ran she would turn and scream giving the performance of her young life before diving to the ground.

'That's wonderful!' exclaimed Gordon. 'Cut! That's a wrap alright.'

He congratulated the actress. 'Well done you!'

'But I was really scared!' she told him, still shaking.

'Did you really think he was going to hit you?'

'Yes I did!', replied the actress. 'As I turned to look over my shoulder, the aeroplane was coming right at me – at eye level!'

Following the completion of filming, it was clear to Gordon that his instincts to use Robin were well founded; that Robin was not a man to take chances; that he would plan meticulously in his calculation of speed and distance whether piloting a car or an aeroplane. Later, when Gordon was offered the opportunity to make a commercial for a well-known local retailer he approached Robin again.

'The managing director says he needs something spectacular. Their theme is people will do anything to go to Tremletts. Have you got any ideas, Rob?'

'I've got something in mind which we could do on Sunday morning,' said Robin.

And so at Plymouth Airport with a view to flying out over Plymouth Sound they loaded up the Auster with Gordon's camera equipment. The side window propped-open provided an ideal aperture for Gordon to operate a hand held 16mm Bolex camera.

'Are you ready?' asked Robin.

'Ready!' replied Gordon.

'You just keep the camera rolling and leave the rest to me.'

Flying low above the water Robin aimed for Smeaton's Tower, the old Lighthouse on Plymouth Hoe. Lifting the nose he climbed the Auster over the tower before banking away then dropping again to pick up the Liskeard Road out of Plymouth. Throttling back to almost a stall and flying as low as he dare, motoring ahead of them was a bus; its back end posted – coincidently and most fortuitously – with an advert that reads: *You are following the bus to Tremletts.*

Robin's low level exploits over the Hoe and the City didn't go unnoticed and he was reprimanded for not having the necessary permissions, but his licence was never in jeopardy and it was just one of those boundaries that Robin enjoyed pushing at whilst getting a very satisfactory result in the process.

Old Fashioned Arrangements

In the early seventies, Robin was still living in digs in Whittington Street, Plymouth with his landlady. Robin, leading such a full life working as a car salesman, flying the Auster, drumming in the band and dating Wendy, probably didn't miss the lack of bathroom or indoor toilet as there was little time for home comforts. Despite the old fashioned arrangements, his devoted landlady looked after him as if he were her son, doing his washing and ironing and making sure that he had something to eat at work.

Robin and Wendy were not prepared to rush into marriage and there were times when it looked as if the relationship was off completely. However, they naturally gravitated toward one another and after four years of courting, marriage was mentioned in conversation one evening over a fish supper – fish that Robin had carefully cooked.

Six months later on May 4th 1973, they married in Plymouth where they continued to lodge in Whittington Street before eventually settling in nearby Plympton. Now with his own premises that included a small garage and driveway to operate from, he could resign from his old car sales job and set up business selling cars freelance.

The Air Race

Onboard the Auster flying over the Midlands, Wendy Bowes – partly to take her mind off her cramped, noisy surroundings, took out a packet of Polos, took one for herself and was about to offer one to her aviator husband – then thought better of it: *What if he chokes?* she thought. Below, they could make out the circuit markers for the forthcoming air race at Halfpenny Green. The air race was to be followed by a Concors d'elegance event plus dinner and dance. Robin had explained to her about the circuits, about how each pilot would be given a handicap because all sorts of aircraft would be taking part, but it did nothing to ease Wendy's apprehension and claustrophobia. She just wanted to be back on terra firma as quickly as possible.

They'd set off from Roborough that morning and were close to making an approach when Robin radioed Halfpenny Green. The Auster was seriously low on fuel and it looked as if they were going to have to put down as soon as possible. He turned to Wendy:

'Uhm, don't panic, don't worry, I'm going to look for a field we're running a little bit short on petrol.'

The Auster was made for short take off and landings not that that made Wendy feel any better as they scanned the area below for a suitable landing site.

'We're going to have to land soon while we've still got a choice.'

They could see some people playing tennis next to what looked like a house and a recently harvested cornfield. Of greater concern was the proximity of towering national grid pylons that would make a field landing very tricky indeed.

'I'm going to go in for a closer look,' said Robin. It looked the best option by far, but always cautious, Robin wanted to be absolutely sure. There was no wind, the field sloped slightly but if it was uneven or littered with stones and molehills the aeroplane could turn over nose first on landing. 'I'm going in a bit lower to look, we've got to land.'

Alerted by the change in engine pitch, the tennis players had by now stopped their games and were looking up at the hazy sky watching intently as the little aeroplane circled their courts. A woman in a nearby house ran out into the garden to shepherd her children off the lawn and out of harms way while Wendy watched the panic below silently contemplating the imminent reality of a crash.

'We're going in!' Robin had made up his mind, but Wendy stricken with fear was unable to affirm or reply in any respect. She was convinced this was a do or die moment.

Skilfully, Robin landed the Auster in the stubble field with nothing more than a bump-se-daisy and kept it down quickly bringing the aeroplane to a short stop well within the confines of the hedges. There was a slight ridge running the width of the field and from that they got a bump on landing but they were down in one piece and that was all that mattered.

Had the landing been part of the competition he'd have earned considerable points. The tennis players dropped their rackets and ran to assist. 'Are you Okay?' one of them asked on reaching the plane. 'We never have any bother like this on a Sunday morning.' At least he hadn't had to put down on the pristine courts.

Meanwhile, Roger Wilkin and his wife Anne, were waiting at Halfpenny Green for Robin to land. Roger was to be his co-pilot for the competition and had driven up from his base at Lynton-on-Ouse. When Robin had failed to show for his expected time and no one at Halfpenny had heard anything Roger began to fear the worse.

'Where have you been?' he asked his co-pilot to-be who arrived chauffeur driven by one of the tennis players and without a discernable aeroplane within which to fly.

'Low on fuel, we had to put down.' Robin opened his ordnance survey map. 'This is my position in a field nearby about two or three miles away. Come with me – I need your advice. And we'll need to pick up some fuel.'

On reaching the field, the two pilots discussed the possibilities of being able to fly out of the field especially in light of the power lines suspended between the towering pylons.

'The way I see it,' said Roger, 'we take the Auster right to the far edge of the field, gun the throttle and either pull up to clear the lines or keep the nose down and fly under the whole lot. At least there's no wind.'

To her relief, with Roger in the co-pilot's seat, Wendy could stand down and watch from the safety of the field, not that that allayed her concerns. By now the police were also in attendance, not that any laws had been broken, but a catastrophe was still possible.

Tennis players and pilots manhandled the Auster as far back to the field's edge as they could get it. Topped up with enough fuel to make the hop to Halfpenny Green, Robin and Roger fired up the engine whilst the police urged the crowd to keep a safe distance. With everyone clear, Robin opened up the throttle and the Auster quickly picked up speed. The small ridge that had given them a bump on the way in would now act as a ramp for lift. On the ground, tennis games temporarily forgotten, the crowd cheered and clapped as the aeroplane lifted into the air, but would it make the power lines ahead? Wendy held her breath.

Keeping the stick forward Robin kept the Auster low preferring to go under the lines before climbing steeply. With take off a success, the power lines behind them, all seemed well, but then to Wendy's utter astonishment and with the crowd watching his every move, he turned the Auster, levelled out then put the plane into a Stuka like dive for a low level pass over the enthralled crowd causing Wendy's knees to buckle as she fainted.

Safe Arrival

At Halfpenny Green with registration for the race completed, the various entrants were taking off in order of handicap according to engine power. Robin and Roger taxied forward under the guidance of a race Marshall. They were optimistic; it looked as if this was going to be a great race with plenty of aircraft taking part and fine weather.

Obediently following the directions from the Marshall, Robin had no way of knowing that there was soft ground ahead until it was too late. The Auster's wheel dug into the soft turf plunging the nose into the ground whereupon the prop snapped and consequently shock loaded the engine. Within minutes of taxiing, their race was over before it had begun.

Despite representations to the race organisers at Halfpenny Green and a subsequent engineering investigation, Robin was never compensated for the loss despite his obeying the directions of a race Marshall. He had no choice but to abandon the Auster at Halfpenny Green until it was repaired at his expense. What with extra journey costs to get home and later return to fly the aeroplane back to Roborough, the Halfpenny Green Race was a costly exercise for the enterprising young pilot. Not surprisingly, he didn't ever enter again.

Death of a Prince

That same year, 1972, Halfpenny Green hosted the Goodyear International Air Trophy. Of its amateur competitors, licensed pilot and President of the British Light Aviation Centre, Prince William of Gloucester – son of the Duke of Gloucester and paternal cousin of Queen Elizabeth II – was a talented pilot with a wealth of air show race experience behind him. Together with co-pilot Commander Vyrell Mitchell registered as "passenger" the duo took off as scheduled in a Piper Cherokee Arrow. The 30 year-old bachelor Prince, ninth in line to the throne, had begun flying as an undergraduate with Cambridge University Flying Club in a de-Havilland Chipmunk of the Queen's Flight.

His passion for flying – both aeroplanes and hot air balloons – might have seemed to those who didn't know him, that his aviation interests were purely frivolous and indulgent: hobbies for a rich young man to while away the years until such time as an official role could be allocated that incorporated royal duties and his hereditary place in the House of Lords. However, those closest to him knew his qualities and dearly loved him for such. Far from being an "air-head" his desire to contribute to society through diplomacy and his sheer energetic enthusiasm were winning him plaudits. He was already making a name for himself as a man of causes: race relations was one such interest. His love of flying was not purely the self-centred devotion of a bored young royal with too much time and money on his hands but rather an amateur occupation that with experience in all its various outlets could lead him to a true understanding of Civil Aviation for which he could act as an ambassador.

It was his friends who were to become his biographers, and it is their contributions that highlight the qualities that Prince William – quietly and without fuss – brought to those relationships. He was a devoted friend, loyal and considerate; a man who did not and would not shy aware from stigma. When a Cambridge University friend became mentally ill, the Prince remained his staunch friend – visiting him in hospital and bringing comfort until such time as the friend passed away. He was a keen advocate of younger generations, encouraging their interests in sport and activities; he was also a man of faith but not to the point of wearing it on his sleeve. Always modest, his faith was quietly kept. Handsome, admired throughout the world, with his company much sought after, his devotion to one woman in particular, Hungarian born Zsuzsi Starkloff, perhaps said more about his loyalty than anything he did in his short but eventful life.

He knew well enough the perils and dangers of flying. Born in 1941, as a baby his pram had been rocked by the explosions of those returning bombers that fell short of their home base in East Anglia. He had first taken to the skies at White Waltham in May 1960 gaining his PPL following twenty-seven hours, five of which were solo. By the time he completed his course with the University Air Squadron in June 1963, his logbook totalled over thirty-six hours – enabling him to hold a PPL; but it was in African skies that much of his early experience was gained and again later in Japan in his role as a trade envoy. In respect of Africa and most remarkably Japan, he would fly the entire route in his own aircraft together with his chosen "safety pilot".

At home in the UK, he was introduced to the various aspects of sporting aviation: gliding, parachuting and ballooning through his chairmanship of the newly formed Aviation Council of the Royal Aero Club. Indeed, he became so interested in British airfields and the history of flying clubs that he set out to meet as many members as he could and the types of aircraft they flew – vintage and modern including privately owned jets.

At Halfpenny Green, the competition into which he entered required that two aircraft with the same handicap rating would start side-by-side for a 'paired' take-off. For this race there were no less that seven paired take-offs in total.

His co-pilot, Commander Vyrell Mitchell, was a highly experienced former Fleet Air Arm pilot. In his civilian role Mitchell worked as a Sales Manager for the Piper company, therefore his tacit knowledge of the type and aviation in general was highly regarded. Further, his association with the prince as a flying partner covered many tens of hours in both flight time and preparation. Together, they had flown a long haul flight in a Twin Comanche to Japan via the Middle East.

By the time of the race, far from being strangers or new acquaintances, a strong bond had formed between the two men.

With the culmination of Prince William's duty in Japan, his hopes dashed of piloting himself home to the UK via the Pacific and the US, he sold the Comanche (the only non-Japanese registered aircraft in Japan at that time) and, on his return home, purchased the single-engined Piper Arrow (G-AYPW) with retractable under-carriage; the aeroplane that he and Mitchell would use for the race.

Alongside them at the start, another Piper Arrow (EI-AVH) owned and piloted by Irishman Tim Philips. Their allocated starting positions were Prince William's aircraft to the left and Tim Philips to the right; their first objective on take-off to pass the "scatter point" pylon before commencing a turn to the left of about 120 degrees that would take them out on their first leg of the course. As Philips was the "right-hand man" it had been agreed between both pilots that it would be his responsibility to keep a clear, safe distance of the prince's aircraft to his left as was normal procedure for that type of take-off necessitating a left turn in the climb.

On the drop of the flag, Philips' aircraft pulling slightly ahead on acceleration to take-off and increasing when airborne had the advantage on the prince and Mitchell now falling behind as undercarriages on both aircraft were retracted. Looking over his shoulder to see the prince begin his turn, Philips began his own turn safe in the knowledge that he wasn't cutting-across his fellow competitor and that he'd achieved the necessary 105 mph to execute the manoeuvre. Now the prince's aeroplane – though still on the inside – was engaging in a very steep turn and at such a low level that it was passing underneath the Irish aeroplane.

That he seemed to throw his aeroplane into such a steep turn so soon after take-off baffled not only those amongst the crowd of 30,000 who witnessed the tragic event but all those who knew Prince William's great caution when flying. Throughout his flying career, when fellow pilots had suggested that he test the handling of aircraft with steep turns such suggestions were always politely rebuffed by the prince. He was not a "press-on" pilot; that was not his reputation. To the contrary he was a highly meticulous and above all a truly "safe pair of hands". Nor would Commander Vyrell Mitchell with all his considerable experience have encouraged such a manoeuvre.

At first hitting a tree and then impacting with a hedge the yellow and white piper Cherokee Arrow crashed beside a lane where three courageous boys made a determined attempt to rescue the men trapped inside – the identities of the occupants unknown to them. Beaten back by the intense heat of raging flames they valiantly tried to break open the doors but unable to do so they tried pulling the broken fuselage in half; sadly, to no avail, but no team of grown men could have tried harder or achieved more than these brave lads who risked their own lives to save others. By the time the first fire appliance arrived, the fire was too great and they were unable to perform a rescue.

Following news of the crash and the loss of the prince, the Queen, the Duke of Edinburgh and Princess Anne cancelled their plans to attend the Munich Olympics. Prime Minister Edward Heath sent messages of condolence to Her Majesty and the Duke and Duchess of Gloucester. In light of his severe illness, the Duchess held news of her son's death from the Duke, though she feared that he might hear of it through television news.

Over the coming weeks and months of that summer, the Duchess received letters of condolence from across the world and she diligently answered all in her own hand. One from Air Chief Marshall Sir Angus Walker encapsulated perfectly Prince William's enthusiasm for flying: "[his] aptitude for flying, and how he applied himself unsparingly to his training during the long vacation at the end of his first year at Cambridge. I have followed with admiration his subsequent career. What commendable initiative and skill he displayed in piloting his own aircraft to West Africa and subsequently to Japan: two outstandingly courageous achievements. The private pilots' centre has lost an exceptional practising champion, and the country one of the most modest and natural members of the rising generation."

Captain John Robathan also wrote in sympathy to the Duchess expressing condolences of both himself and his wife: "… [But] nevertheless the recollection of those concerned survives the passage of the years, and they remain remembered as the gay, happy and adventurous people they always were."

It has been said that Prince William's real freedom was to be found in the sky. Admitting to some dangerous moments in the air, he said: 'There were a few times when I wondered what was going to happen but I found this exhilarating and never thought of crashing. You simply can't fly with fear.'

Insisting that he would have gone ahead had anyone tried to forbid him from flying, he said:

'You don't know what it is like to be so divorced from the earth. I feel another person when I am behind the controls. I'm never happier than when I have a weekend clear and can take off with friends. It's my only real relaxation. I need the freedom of flying. Thank goodness no one ever tried to stop me. If you are Royal, everybody gets scared of you taking risks.'

Adventurous People and the Rising Generation

Back home in Plymouth, spot landing competitions at Roborough were fun and far less costly. The objective was to touch down closest to a particular mark or line. A natural flyer, Robin was a frequent winner.

Plymouth aerodrome in those days was just a grass runway with a cabin and control tower both constructed of timber and used by the flying club and Britannia – the Royal Naval College at Dartmouth. The RN college students did their elementary flying training at Roborough before selection into the Fleet Air Arm if they made the grade. During WW2 it had been a base for spitfires and hurricanes defending Plymouth and its docks from the Luftwaffe's bombing raids. And ironically for Robin, 'Faith' – one of the famous three Gloster Gladiators (Faith, Hope and Charity) that had protected Malta during the war years – was on view by the runway. There was a cafeteria and club offices where Robin's flying instructor Bill Lucas spent much of his time.

When Roborough became the operations base for Brymon Aviation, the landing prices went up. For club members the days of spot landing competitions were over, and for novice pilots such practise provided invaluable experience. Now, there would be a charge for every touch and go landing. It wasn't long before the club decamped from Roborough and set up a new base at Cardinham on the edge of Bodmin Moor.

Home for Mr and Mrs Robin Bowes was in the pleasant modern suburbs of Plympton. Wendy had been working for Westward Television for ten years and Robin was supplementing his car sales job with regular drumming work in the evenings meaning that the newly-weds passed like ships in the night until Wendy came off the late shifts and worked 9am to 5pm five days a week and every other weekend before being promoted to a management PA.

Chapter 5 – And There it Lay

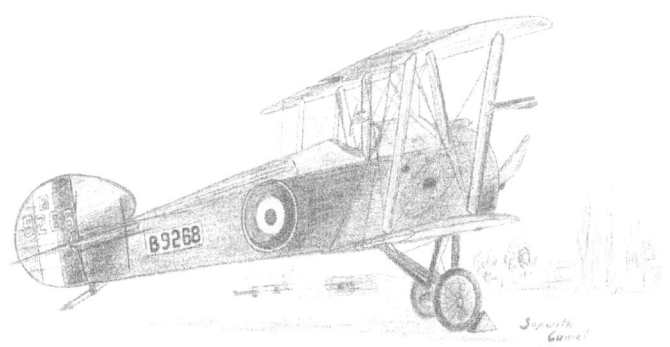

Line drawing of a Sopwith Camel by Robin Bowes

"Flying my triplane for the first time, I attacked with four of my gentlemen a very courageously flown English artillery flyer. I approached and fired 20 shots from a distance of 50 metres, whereupon my adversary fell and crashed near Zonnebeke. Most probably the English pilot took me for an English triplane, as the observer stood upright in the aircraft without thinking of making use of his gun."

Baron Manfred von Richthofen: September 1st. 1917:
Fokker F1 triplane, No.102

After a safe landing at Roborough, Wendy lined up Pat and Robin in the viewfinder. The two men stood proudly in front of their new, red, iron crossed acquisition and she pressed the shutter. Click! One for the album – posterity. Proud smiles could not hide their glee at what they considered this most remarkable find. But to Wendy, any amount of smiling could not hide the instantaneous feeling of hate she had for the triplane. From this first moment she loathed it; and yet, to look at Robin, it was as if he had just given birth. She was determined at this moment not to tell him how she felt. He was so proud of it.

Some three months earlier, the two men had quite by

chance stumbled across the triplane abandoned. Nothing more than a forgotten heap in the back of a hangar with only the company of other WW1 replicas at Land's End Aerodrome, St. Just; and as far as Wendy was concerned, lonely and abandoned was just how it should have stayed. From the moment they saw it, full of creative ideas, the two men discussed the triplane incessantly. Wendy, however, was not at all happy. The sight of the red triplane triggered something deep within her that she couldn't quite put her finger on. In her opinion, the plane was trouble – *would be* trouble.

Robin Bowes in the triplane cockpit

December 1983

The plan was this: they would buy both the Fokker triplane Dr1 replica and the Sopwith Camel and take them as flying exhibits to air shows, regattas, county shows, where they would demonstrate them as static displays at presentations and exhibitions. A sponsor could be found; someone who was in need of original promotion and what could be more original than a brace of WW1 aircraft recalling a more chivalrous era in combat flying.

Ideally, the aeroplanes would portray the legends associated with their types. The triplane would be the Red Baron's and therefore scarlet. The Camel would be the olive green of Captain Roy Brown's machine – the last man to engage the Baron in aerial combat and, ironically, one of the first aircraft to fly for the newly formed Royal Air Force. The colour schemes would have to incorporate the sponsors' logo, but with marketing in mind, the exposure by way of advertising and all sorts of publicity material from magazines to show programmes would be worth its weight in gold. Maximum publicity value sprang to mind. Other companies used similar aviation themes, but there were no legendary First World War replicas, so there was an opportunity for sure; a Unique Selling Point.

Both Robin and Pat believed that these aeroplanes would create a far more lasting impression on sponsors and public than a succession of modern characterless aircraft, and they would lend themselves naturally to certain events.

Even covered in dust and de-rigged, Robin considered the triplane by no means a hopeless case. With finance and a few visits, it could be prepared for a test flight in no time at all; and just maybe it was a realistic proposition to buy. The Cessna Aerobat he currently owned could be put up for sale and any additional finance could be found; nothing was beyond the realms of possibility.

As interest in purchasing the triplane grew, Robin looked for any information he could find on the Red Baron.

In the following months, he read everything he could lay his hands on regarding flying in the First World War. So acute became his knowledge, especially concerning the Red Baron, that he could have happily entered the BBC programme Mastermind and sat in the famous black leather chair to answer any amount of pertinent questions from Magnus Magnusson on the life and death of the Baron or on early aviation in general.

The idea of flying in the guise of the Red Baron had immense appeal. After all, the Fokker triplane was a unique piece of aviation history. Three wings as opposed to two, coloured bright red with large black crosses on white backgrounds. His belief being that the triplane would: *make 'em look up from the ice cream and hamburger queues. Who'd ever seen an aircraft like this fly unless it was in a film?* Most Second World War veterans would never have seen such an aeroplane in the sky. The triplane would be a flying history lesson and provide an opportunity to understand just what it was like to fly a type of aircraft that had become such a big part of aviation legend. Neither Robin nor Pat needed much convincing; they were sold on the idea as the plane was surely bound to attract a lot of publicity. For a man who wanted to make his entire living as a display pilot, the concept of flying as the Red Baron appealed like nothing else and might even provide untold opportunities.

Wendy, though not happy with her husband's decision, was determined to put a brave face on it and show willing. She knew well enough that when "Wobbly" determined to see something through, then nothing would stop him or even put him off his stride. Seeing that his mind was made up, she popped into a Plymouth department store and bought him a white silk scarf, an item he cherished and would wear it always when flying. Her gesture, especially in light of her fear and hate of the triplane was magnanimous, but the excessive time Robin was now putting into his flying was beginning to damage their relationship.

With the Aerobat sold and a local finance company providing the remainder of the cash needed to purchase the little red triplane, Robin and Pat secured a deal with Leisure Sport Limited – the plane's owners. It had been built for

Leisure Sport by one of the great characters of British Aviation, Vivian "Viv" Bellamy, a former WW2 Royal Navy pilot and post war aviation entrepreneur and raconteur who had built several WW1 replicas at Lands End including the Fokker Dr.1, built to plans supplied by Walter Redfearn & sons Incorporated, Postfalls, Idaho, USA.

Bellamy had come to Lands End to run the aerodrome after an extraordinary career in aviation. During WW2, he'd served in the Fleet Air Arm flying a variety of carrier borne aircraft before being hospitalised trying to rescue a fellow pilot from a burning Corsair that had crashed on the deck of HMS Illustrious – then Britain's newest and biggest aircraft carrier – whilst in defence of Malta. Bellamy's heroic attempt however, was in vain as the pilot and two other ships crew died in the flames.

After the war, and at different times, he'd been an owner/operator of an airfreight company; he'd restored aircraft and flown a Supermarine Walrus to Timbuctoo before becoming proprietor of the Hampshire Aeroplane Club at Southampton Airport. His passions were collecting and rebuilding prop-engined aircraft in the jet age just when everybody else was casting them off as relics of a bygone era.

He built his first triplane – an Avro – for the film *The Magnificent Men in their Flying Machines*, and when not building aeroplanes or flying, he'd work as a flying instructor. But despite all his hard efforts the Club, though well patronised, did not make money and so he moved to Tunis where he spent much of the sixties as a flying instructor before returning home to work on the film *Battle of Britain*, in which he flew every type of aircraft featured.

Now at Lands End, this larger than life character, with his Royal Navy full set and thinning hair – not unlike the actor James Robertson Justice in both appearance and character – could be found with his faithful Labrador at his side. When

Viv went flying in the club's Cessna, the dog would sit in the front seat alongside his master. The pair were inseparable.

The Fokker triplane was not his only building project: other replica planes receiving the Bellamy treatment were a Sopwith Camel; a D.H.2; a Strutter, and a Fury.

Claude Graham-White

By 1984, Robin had prepared the triplane in readiness for its first flight. The triplane might have been ready, but despite all his homework on WW1 aces and their planes, Robin certainly wasn't. He was about to follow in the pioneering footsteps of the early aviators – men like Claude Grahame-White who flew the very first powered aeroplane over the skies of West Cornwall on 23rd July 1910.

Grahame-White was not only a pioneer aviator; he was something of a prophet, too. Despite the flimsy box kite construction of these early aircraft he had an ability to foresee the strategic fighting potential of the aeroplane in war and with this vision in mind he approached Admiral May of the Royal Navy for permission to drop an object onto the deck of the flagship Dreadnought – anchored in Mounts Bay that day as part of a combined fleet exercise. The idea caused considerable excitement in the press, who were quick to report on the possible offensive capability of a little aeroplane that would be able to avoid the firepower of a ship's big guns as they put it. The idea of military air power came a step closer that day in 1910.

Strong Cornish winds kept Grahame-White's Farman biplane on the ground for most of the day and it wasn't until 6pm that evening that it became calm enough to even consider a take-off. Crowds had gathered at Poniou Meadow and each paid half-a-crown to watch this most historic of events. Their patience was eventually rewarded when just after 8pm the Farman took to the skies for a three-minute circuit of the field at some 200 feet. Witnesses to history, the jubilant crowds thrilled at the spectacle of a motor driven flying machine in their skies and so skilfully flown by a remarkably brave new breed of hero. Just before dusk, Grahame-White took off a second time this time heading west toward the fishing village of Newlyn. Over the harbour he turned east and headed out over the sea where he rendezvoused with 200 mighty warships of the combined fleets of the Royal Navy majestically anchored in Mounts Bay for an informal review by His Majesty King George V. In salute to the miniscule and fragile machine circling above them each ship sounded claxons and whistles much to the delight of the pilot and all those watching from the shore. After approximately 15 minutes and having covered some nine miles, Grahame-White returned to Poniou Meadow to rapturous applause from all those gathered. Later that same year, he successfully completed the Daily Mail London-Manchester Air Race arriving in second place. A truly innovative aviator, he fitted headlamps to the Farman enabling him to take off in the dark.

A flamboyant visionary of working class stock, who'd made his money as a car dealer, Graham-White saw clearly into the future of aviation. But he wasn't simply a visionary or soothsayer: he demonstrated as a pilot with remarkable skills not only what an aeroplane could do as a machine of war, but also as a machine of mass transport. At a time when the great sea going liners held sway on the world's oceans, and Marconi's radio experiments from the cliffs of the Lizard Peninsula were opening up the world to mass communications, Claude Graham-White was at the forefront of the development of civil aviation that he believed would soon be carrying people across the globe.

Despite his humble origins, he successfully influenced the great and the good, the movers and the shakers of the day into understanding that the aeroplane was not simply a flying gimmick of the circus, but a substantial creation that would change the world for good.

By the time of the Great War, Graham-White had established himself as an architect and builder of aerodromes and aircraft production factories where he employed fairly and with great humanity thousands of workers – many of whom were women. They worked, and were even often accommodated, in modern state of the art buildings with good lighting that were to influence architects throughout the twentieth century and beyond. He developed aeroplane types and even set up a school for the training of pilots at premises he'd created at Hendon. He was in many respects, von Richthofen's greatest adversary in respect of the scale of the aeroplanes and pilots that his factories and flying school were producing. A truly practical man, he was not averse to rolling up his sleeves and working alongside his staff on the production lines or in the drawing room. It is reasonable to assume a century later that had it not been for Claude Graham-White's gargantuan efforts in the development of aviation, Great Britain could not have fought on an equal footing against Germany and the Austro-Hungarian Empire in the way it did. Air supremacy would have been with the Kaiser.

Fokker Dr1 triplane replica, G-BEFR.

The first time Robin was airborne with the triplane over the Land's End coast, anxiety kicked in. To his horror, something appeared to be seriously wrong and only a miracle was going to prevent a catastrophe on landing – that's if he could turn and get it down at all. The triplane was literally skidding through the sky and seemingly out of control. Not one to panic, he went over and over in his mind just what he knew about the plane, the history of the type and the history of this replica in particular. What had he missed? Was there anything he hadn't taken into account? *Must be rigged wrongly?* he thought as he struggled to get it to fly in a straight line.

Then it dawned on him that with no fixed fin area, there was little or no directional stability. The whole comma shaped tail plane was the rudder unlike modern aircraft, which are stable enough for the pilot to fly with feet off the pedals using only occasional rudder to co-ordinate turns. Thank goodness! It wasn't rigged incorrectly at all, but boy, it was fighting every effort to control it.

    Once he got used to the feeling that it was perpetually falling over, he thought it more than prudent to land the awkward machine whilst there was still enough fuel for several attempts and made a promise to himself that if he ever managed to get the darn thing back on the ground in one piece, he'd put it into the next auction! He needed as much grass runway as he could find. The aircraft had no tail wheel and is of a type known as a "taildragger" meaning literally that a steel "skid" or runner would act like a rail on a sled when the tail made contact with the ground. To his relief, the all-grass aerodrome allowed him to face the plane into wind. The thought of trying to land this little monster in a crosswind was a thought too far. What he wasn't prepared for was the fact that once he'd gingerly manoeuvred the triplane into a final approach – using every bit of flying skill within him – he could no longer see the strip as the middle wing now obscured his forward view.

Summoning up "blind" courage, feeling his way, he would have to make an assumption that the area ahead was as clear for landing as it had been when he was last able to see it. Worse, he wasn't even in radio contact, so there was no way he could talk to a ground controller. Only by deftly sideslipping the aeroplane could he peer down the side of the fuselage and engine cowling in order to get some vision of the ground to come.

By what seemed to him like a miracle, he put the aircraft down on the strip without further incident and vowed to Pat – and to himself – that he would sell the aircraft at the first opportunity.

As enthusiastic as he had been to buy the triplane and make capital from it through display flying, Robin, characteristically, was aware that it would take time before he had the experience to fly it with confidence. Also, he would need to attain an Air Display Authorisation and to gain that, his routine would have to be scrutinised by representatives of the British Aerobatic Association and the Historic Aircraft Association. The "tripe" as he was to call it, had taught him a valuable lesson in that first flight and he realised that in order to learn more he would have to put in a fair few hours getting to know his new "partner".

Triplane in flight (credit Paul Harrison Photography)

The DH Lawrence Connection

Just a few miles up the coast from Land's End airport is the village of Zennor, briefly once home to the writer DH Lawrence and his German born wife-to-be, Frieda Weekly (neé von Richthofen).

Frieda was the cousin of the man who was to become the Red Baron; her brief tenure in West Cornwall shortened by the growing animosity of locals in that summer of 1914 – just weeks before the outbreak of war.

Their relationship with local people in this most remote part of Cornwall was nothing so simple as fraught. They were suspected and accused of spying and collusion with the enemy. Frieda in particular was accused of running to the cliff tops to send semaphore signals to German submarines cruising close by. Nothing could have been further from the truth: Lawrence was a fervent anti-militarist and wanted nothing more than to concentrate on his work in what should have been a peaceful backwater for a novelist.

Accusations of spying were nothing new to him. Ironically, it had been during the couple's elopement to Germany that the authorities arrested him on charges of spying for Britain while Frieda's aristocrat father, Baron Friedrich von Richthofen, nobly came to his aid and so secured his release.

With War approaching the couple fled Germany and the Continent for England, though they soon longed to return.

During his time in Zennor, Lawrence penned *Women in Love*; however, its frank and forthright explorations of the human condition was considered so desolate and sour in its content that it wasn't published until two years after the war's end in 1920. His most controversial novel, *Lady Chatterley's Lover*, published in 1928, was said to be based on Frieda's experiences as a young aristocrat and her extra marital relationship with the working class Lawrence.

The bitter antagonisms that grew in the Zennor community during the period of the First World War meant that the couple found themselves harassed not only by locals but also by the British authorities. Like all creative artists, his experiences however dramatic were grist to the writer's mill, and he would later incorporate some of his more severe Cornish experiences in fringe politics in his Australian set novel *Kangaroo*.

By late 1917, under the terms of the Defence of the Realm Act, the couple were forced to leave Zennor and Cornwall for good, initially finding refuge in rural Berkshire and later Derbyshire. The couple left England for good in November 1919.

Going Against the Norm

At Bodmin airfield situated near the busy A30 at Cardinham on the edge of Bodmin Moor, Robin set about putting into practice a display routine with the triplane – much to the amusement of the members of the Cornwall Flying Club. They even forgave him for flying against the normal circuit pattern and their presence on summer days that year watching from the grass outside the club house encouraged him to persevere. Not that he needed much encouragement. He would know when it was time to bring the rehearsal to an end when he could see his "audience" traipse back inside the clubhouse to attend to liquid matters at the bar. Many were old friends from Roborough days – exiles like him – displaced and bereft of their club HQ when Plymouth became a commercial airport and Bodmin had become their new base.

They also primed him for the many questions that would come from larger audiences on the Air Show circuit.

'What's the triplane like to fly? Was it made for a film? Is it full size?'

Generally, they didn't prod it as if it were a caged animal in the way that air show crowds were prone to do. Being the focus of attention could be a disadvantage, though. He was realising that every time he took off and landed, all eyes were on him. This attention would be magnified several thousand times at each venue once he began the real display circuit.

Learning to fly in the Red Baron's seat was giving him a taste for flying as it would have been in the years of the First War. Then, airfields were literally that – no runways or taxiways. Face the aeroplane into wind, run up and take off hopefully before reaching the hedge. Similarly, landing had a different set of skills to be mastered. Cross the hedge into wind, try to catch sight of something familiar between the wings and with no brakes, learn to coast to a halt. Trying to taxi on the ground was especially difficult. Nose up, tail on the ground, the middle wing prevented a view of any kind; the experience was not unlike trying to park a car with the view obscured by a raised bonnet. The only way to try and see ahead was to peer around the fuselage and engine cowling. Landing with such poor visibility required side slipping a little to get a view and hoping that as the field was clear the last moment you looked, it would remain so for touch down.

Reinhold Platz – Fokker's chief designer – was aware of the problem and thoughtfully cut out sections at the wing roots of the middle wing, but it only increased forward view by a few metres. Robin was particularly concerned because on the approach he was unable to see the horizon making it difficult to judge whether the wings were level or whether there was a serious drift. Amazing that a pilot in the last quarter of the twentieth century was experiencing the same problems as his predecessors in the first quarter. On the ground, to change direction without a wing walker, it was theoretically possible to blow the tail around using enough power and rudder, but without brakes each burst of power simply made the aeroplane charge forward with little inclination to turn. Without a wing walker, the alternative was to unstrap, clamber down and lift the tail much to the amusement of anyone watching whilst the engine obediently idled. If it didn't, then the most Chaplinesque routine would be underway as novice triplane pilot chased errant aeroplane. If such a thing ever happened in those salad days, he never told anyone.

Although the triplane was a replica, the engine was itself something of a war veteran – not from the First, but from the Second War: a 165 h.p. Warner Scarab radial originally from a Fairchild Argus. The engine had a real history in this respect and had it talked, could undoubtedly have told some stories. In straight and level flight the engine powered the triplane along at a respectable 85 mph burning nine gallons of aviation fuel an hour. The rate of climb was actually inferior to the original aeroplane at 600 feet a minute, about half of what a triplane could achieve in 1917. Robin put this down to very coarse pitch prop. Also, the aeroplane weighed 400lbs heavier than the original, even though the dimensions were the same.

The triplane was even equipped with two lethal looking Spandau machine guns mounted on top of the fuselage just forward of the pilot's cockpit. These would keep the weekend Cessna flyers in check at the Flying Club! The replica guns even had the stamp markings and serial numbers of those actually removed as souvenirs from von Richthofen's doomed aeroplane no.425/17.

The colour was that favoured by von Richthofen – the Red Baron. Odd, now, that in a world where camouflaged warfare was being premiered for the first time, the Baron had chosen the most conspicuous of colours. His choice at first glance appears to reflect an aristocratic arrogance, harking back to an era of scarlet tunics and cavalry charges, and – perhaps considering the Baron's earlier military career – there's something in that, but there are other theories.

The Baron would lead his "Jasta" (Wing) and it would be

fair to assume that in an age where radio communications between aircraft was not possible, visual signs were the only way of making your order known. Gesticulations play a part in this if the flying is slow, close and careful, but if in the milieu you want to concentrate on your leader, you need to be able to spot him. Of course, the enemy too, wants to pick off the leader; but in a highly manoeuvrable aeroplane such as the triplane, it's a case of catch-me-if-you-can. The leader has the experience, the faster engine, the best guns, the most modern design in the squadron, so for all those advantages, sticking out a bit is probably not such a high price to pay. Camouflage was being used in war for the first time during this period. Aeroplanes on both sides used various colours for camouflage from greys and blacks to olive green. It's tempting to speculate, too, that a scarlet aeroplane is less easy to see coming out of the sun.

Von Richthofen flew a small number of triplanes and reputedly painted each red to a greater or lesser degree, though they were delivered in standard camouflage before being painted over in red dope, which also covered fuselage markings. The painting was quite indiscriminate.

In painting his aeroplane red, the Baron was probably making a statement for the benefit of his own people as much as for the benefit of the allied air forces. Such individuality was not known on the other side of the Western Front and archive pictures of early squadrons show uniform ranks of British and French aircraft – a tradition carried into WW2. It was obvious that the Baron wanted to be recognised. He could hardly flash his Blue Max medal at a downed aviator several hundred feet below him, yet from an early stage in his flying career he seemed determined that his successes be noticed. The collection of souvenirs including the commission of special drinking cups – at great personal expense – and the lavish decoration of his room in "trophies" all suggest an obvious desire to be noticed. Yet he was notoriously quiet and shy and politely avoided the excesses of the officers' mess wherever possible. It was as if von Richthofen was consciously creating his own legend.

This still doesn't explain why red as opposed to any other colour. The answer might be found in von Richthofen's first career as a cavalry officer in the Uhlan Regiment Nr.1 "Kaiser Alexander III". By tradition, red piping and banding made up the regiment's uniforms including caps. If anything, the colour says more about the power and influence of the man and his unique position in German society in the last two years of his short, yet remarkable life.

Some commentators have suggested that von Richthofen was never happy with the various camouflage paint schemes that were being imposed from high command and that he painted his machine red in defiance.

Whatever the truth of it, the colour scheme also worked in Robin's favour. If he was to attract sponsors and catch the eye of the crowds, there was no better colour than red. Without doubt, he would be noticed. Air show organisers were most certainly interested in the Red Baron.

Triplane on grass (Paul Harrison Photography)

Off the drawing board

Walter W. Redfern had not originally considered making his carefully researched plans available to a wider market, but growing interest in his triplane replica caused him to reconsider. This was a business opportunity that couldn't be easily ignored and so Redfern decided to make the plans available.

His triplane – or "ship" as he called it in line with American aviation terminology – was the same height, length and span as the original WWI type, though heavier, surprisingly, considering the lighter alloys of the late twentieth century. Centre of gravity was the same, but Redfern considered his plane stronger than the original.

In describing its flying characteristics, he wrote:

'The tripe is not hard to fly but for better ground handling, I would use a steerable tail wheel and spring. My first ship has a skid with a small wheel in the end of it. You have to fly the tail around on the ground. When you fly the ship for the first time, wheel it on until you get the feel of it.'

Triplanes: the First Pioneer

The story of the triplane type predates both Sopwith and Fokker, and even the Wright Brothers first flight.

The very first recorded triplane was constructed by one Percy Sinclair Pilcher; a pale yet determined man who may well have beaten the Wright Brothers into the air with an attempt at powered flight on September 30th, 1899. Unfortunately, he was killed in the attempt at flying a glider he called the Hawk.

On that same field at Stanford Hall, the home of his friend, Lord Braye, stood a remarkable, innovatory machine – a triplane with which Pilcher was going to attempt powered flight. Together with his engineering partner, he had manufactured internal combustion engines; it was believed by witnesses that the triplane was due to be powered into the air by a Wilson-Pilcher engine that may well have produced four horse power – maybe even five. Word was that the engine had proved troublesome when Pilcher tried to increase its power and consequently broke the crankshaft. People speculated long after the event as to whether the engine was ready that day, but what is known is that a large crowd had assembled at Stanford Hall for something very special; among their number was Lord Baden-Powell.

Pilcher was something of an experienced glider pioneer, having made several unpowered flights reaching, on occasion, 400 yards at heights of between 20 to 50 feet during trials in the mid-1890s. He got his gliders into the air just as hang glider pilots do today by running down a slope and into wind. He wasn't alone in his attempts: in Germany Lilienthal strove for "aerial balance" through the mastery of gliding, but he was killed striving for that mastery in 1896 after 2,000 flights.

Powered flights were restricted to small, steam powered model aeroplanes and due to the immense secrecy that surrounded such attempts the rest of the world often didn't know just what had been achieved until some years later. Indeed, that was also the case with the flight of the Wright brothers as it took time for the word to spread and for people to believe what had actually happened.

Welshman, William Frost, took out a patent in 1894 for "A flying machine", but he was cheated by fate when the machine was destroyed on the ground by a fierce storm. A builder by trade, he lacked the finance to rebuild his dream and offered it to the War Office who, with characteristic lack of foresight, returned his offer with the words: "This nation does not intend to adopt aerial navigation as a means of warfare."

Percy Pilcher was a name that could have topped the list of aviation history pioneers had his triplane powered into the sky that day. It remains an irony that he decided to fly his Hawk glider first. At 4.30 that afternoon, Percy took off and the glider lifted to an estimated 30 feet when a hemp rope securing the fuselage broke causing the tail to collapse. The Hawk pitched into a forward somersault and the wings folded resulting in a crash that took the life of a man who may just otherwise have made the world's first powered flight – and in a triplane.

The triplane remained at Stanford Hall for some time until it was dismantled, its parts given out to members of the Pilcher family as keepsakes of a courageous man and his inventions. Today, one complete wing panel survives, but the mysterious engine vanished without trace. A quarter scale version of Pilcher's triplane flew many generations later with a light, modern engine, proving that its design was good. And his drawings were taken to Kitty Hawk, where, four years later, they proved invaluable to the Wright brothers.

Wilson, his business partner, and with whom Pilcher had built a car, went on to make a fortune from the gearboxes he supplied to Rolls Royce. Today, in the English Midlands, an obelisk stands to mark the spot where England's first aviation casualty fell.

VIP Status

The triplane ensured Robin's VIP status at every event. In short, it was eye-catching – a show stealer. Even in the company of genuine vintage aeroplanes – that in their respective and extensive histories had either fought in battles or circumnavigated the globe piloted by some of the most notable aviators of the twentieth century – the little triplane replica was assured of write-ups and glossy pictures in magazine features.

It wasn't just air shows where the triplane would make an impact. By the close of the flying season in October 1985, Robin and his triplane were the sole representatives of WW1 aviation at a V.A.C. Spot Landing Competition at Popham airfield, Hampshire. Sited alongside the busy A303 London to Exeter road, Popham as a home to weekend flyers is always a distraction for motorists as the runway is in clear sight from both carriageways. What accidents or near-accidents were caused on the 303 that autumn Sunday by the spectacle of the Red Baron looping in Hampshire skies was probably never recorded.

The visit merited considerable interest from all who attended and a paragraph with picture in *Vintage News* the magazine of the Vintage Aircraft Club. That first winter on the circuit, the triplane was securely stored with the Historic Aircraft Collection at RAF St.Athan, South Wales.

At Christmas friends and family all received triplane cards specially designed and handmade by Robin and Wendy with Santa Claus on his annual round the globe trek, but this time at the controls of the triplane with not a reindeer in sight. Even Wendy received one.

It wasn't all Plain Sailing

There were problems that would dog Robin and the triplane and here, Gordon Clarke, who'd recognised the triplane from its days at Thorpe Leisure Park, recalls one incident that could have become very serious.

"So with Pat Crawford flying the Cessna, we rendezvoused over Dartmouth – he'd been doing a show – and we did the air-to-air stuff.' Then Robin said he had to go back to land. And he got down very quickly. What had happened was, on the rudder section one of the pivots was actually broken and he had about fifteen minutes left to get in. So they rushed it down to St.Just to have a new hinge made and fitted. From then on, the plane was fine."

Chapter 6 – Airymouse Flies Again

John Robert Currie's "Wot"

Ever since man first flew there has been a quest for a cheap to run affordable aeroplane that could be flown by the average enthusiast and not just a privileged minority. With this object in view, the Lympne Trials were arranged in 1923. Henri Mignet gave the world his "Flying Flea" in the thirties; Mr.Piper came up with the "Cub", and de Havilland his "Moths".

Many other lesser known names pursued the same goal and a former RFC pilot, John Robert Currie, produced his design for an affordable aeroplane in the mid-thirties. He built two airframes of his "Wot", as it became known, constructed at Lympne, Kent, by students from Chelsea College of Aeronautics.

Registered G-AFCG and G-AFDS, Currie's Wots made their debut in 1937. Both aeroplanes shared the same JAP JA99 35 hp engine, but were destroyed in an air raid when the hangar in which they were stored was bombed in 1940.

Eighteen years later, Currie was employed as Ground Engineer at the Hampshire Aero Club, then run at Eastleigh Aerodrome by Viv Bellamy. The quest for a cheap-to-run, affordable aeroplane was still on. Like most clubs at that time, Hampshire A.C. were operating Tiger Moths, Austers and Magisters, and Currie was persuaded to revive his Wot plans assisted by John Isaacs and funded by Viv Bellamy.

The first aeroplane, G-APNT, was constructed in seven months making its first flight in August 1958; powered by a similar JAP 35 hp engine that had been fitted to the two pre-war machines.

Contemporary air tests described the handling as "sheer

delight" and with the aircraft available for £600 ready to fly away, sets of plans were made available at £10 each for the amateur builder for whom it would be an economic aeroplane.

However, the search for higher performance soon saw the trial installation of a Walter Micron II of approximately 60 hp. In this configuration the aeroplane was nicknamed the "Hot Wot" and it showed great promise. Experiments were even made on the Hamble River with floats fitted. This version was known as the "Wet Wot".

By autumn 1959 the second Hampshire Aero Club aircraft was completed as G-APWT, initially with the Walter Micron II engine, but after more float trials an even more exciting development took place with the installation of a Rover TP60/1 turboprop engine. Inevitably this became known at the "Jet Wot", but reverted to Walter Micron power again by 1962 as the "Hotter Wot" with the Micron III 65 hp engine.

In this form during 1960 it took part in the prestigious International Lockheed Acrobatic Trophy competition flown by Flt Lt H R Lane, which though nimble could not quite match the performance of the specialised aerobatic machines.

Meanwhile a third airframe was under construction by Dr J H B Urmston who had flown the Hampshire Aero Club's Wot and promptly fell in love with the design and its flying characteristics. Urmston's humorous description of building the aircraft G-ARZW, was immortalised in his book *Birds and Fools Fly*, which became a must-have companion for anyone remotely interested in light aircraft at that time.

By now word had spread of the delightful little biplane and more amateur builders were starting work on the design using Continental, Lycoming and Pobjoy engines as well as the Walter Micron and JAP power plants. So whilst the Wot had not gone into full production as had been hoped, the faith and enthusiasm in the design shown by Messrs Currie, Bellamy and Isaacs had not been wasted.

Meanwhile the first production Wot: G-APNT started its second career and was purchased in 1959 by one, Harald Penrose.

Enter Harald Penrose

Born in Hereford on April 12, 1904, not long after the Kitty Hawk had made its maiden flight from the oddly named Kill Devil Hill, Harald James Penrose was a true child of the new aviation era. Aged five, he stared enthralled at a picture of Louis Blériot's monoplane that had successfully flown the Channel. Two years later, he was lifted into the air by a kite he was flying in the park.

No opportunity was missed in persuading his parents and grandparents to take him to Hendon where he would gaze with delight at his heroes in their flying machines. The young Penrose would stand witness to anything remotely connected with flying and every spare moment of holiday he would cycle to watch not only the flying, but also the mechanics tightening bracing wires and tuning engines.

At Reading in 1919, an aunt took him to see Alan Cobham's flying circus in which he watched death defying stunts performed by first war veterans flying their redundant fighters; biplanes such as the Avro 504K and in which aunty treated him to a two guinea joy ride with Cobham himself. The flight marked a watershed in his life: from that point on he knew that he could be nothing other than be an aviator.

Geoffrey de Havilland, then building and flying aircraft of his own designs, was the epitome of all Penrose aspired to be. And so in 1922 he began an aeronautical engineering course at the Northampton Engineering College of London University after which he wasted no time in joining Westland Aircraft at Yeovil as a designer-technician before learning to fly at the Bristol Flying School as a commissioned officer in the Reserve of Air Force Officers in 1927.

Pilot Officer Penrose returned to Westland's where he was appointed manager of the company's Civil Aircraft department in 1928. He flew a considerable number of new machines, overseeing the development of innovatory types such as the W.IV tri-motor, Wessex and Wapiti. When in 1931, the company's chief test pilot, Captain Louis Paget was crippled in a crash, Penrose was appointed his successor as well as being elected the youngest ever Fellow of the Royal Aeronautical Society.

His testing over the next thirty odd years involved flying an incredible diversification of types including the Pterodactyl – an aptly named tail-less monoplane; the altitude Wallace; the tandem wing Lysander; the Houston-Westland PV3 which flew over Everest in 1933, and the P7 – an enclosed cockpit monoplane from which Penrose unintentionally tested the suitability of baling out through a side window when the main canopy refused to open due to suction forces.

In 1935, he designed and built his own Pegasus sailplane, but the relative calm of the period would soon be shattered by the outbreak of war. From 1939, Penrose's abilities were in constant demand testing not only Westland's own aircraft but Spitfires, Seafires, Barracudas, Curtis Mohawks and Tomahawks – all being built under licence at the Yeovil factory.

After the war ended, he was awarded an OBE for services to aviation and survived six years test flying the notorious Wyvern fighter for the Navy which was to claim the lives of three of his fellow test pilots and almost killed Penrose when an aileron linkage broke causing the aircraft to invert at low altitude. The post war years also saw him pioneering the development of the revolutionary autogyros and helicopters that were heralding a new era in aviation history.

In 1953, he was appointed Sales Manager and whilst there was still plenty of flying to be done, his desire to recapture and re-awaken the romance of those earlier days of pre-war flight were answered when he purchased Currie Wot G-APNT and promptly named it *Airymouse*.

The next few years were happily spent flying *Airymouse* over the counties of Wessex. He would fly over the house he designed in Nether Compton, Dorset, where upon hearing its engine, his wife Norn would step out into the garden and wave at the little red biplane framed against blue summer skies. On these unhurried trips of self-discovery, he was able to mix his twin loves of aviation and ornithology. His love of the Wessex countryside as seen from the cockpit of G-APNT was immortalised in the book *Airymouse*, published in 1967. No stranger to literature, he was a prolific writer and wrote his first book, *I flew with the Birds*, in 1949, and the classic *No Echoes in the Sky* in 1958. When he wrote the biography of aircraft designer, Roy Chadwick (*Architect of Wings*), and included in the opening chapter an incident in which Chadwick as a boy is inspired by the sight of another boy flying a kite; an incident that: '… must have been the awakening of aeronautical consciousness', he wrote. One can't help but think that Penrose was also drawing on his own experience of "awakening"; that of being lifted into the air by a kite. The splendid ability of Harald Penrose to communicate his knowledge and love of nature and flight; his gift of being able to expound on subjects so dear to him enthralled many an aviation enthusiast – among them the impecunious Robin Bowes.

Following a long absence, Airymouse – reassembled and tested – returns to Westcountry skies

Peering into hangars

When Robin and Pat peer into the dim hangar and see a Currie Wot sitting cheekily amongst several sleek modern sailplanes, they know that it will be impossible to resist the character and charm of such a trim little biplane.

They are certainly not the first enthusiasts to search for an aeroplane that is economical to operate yet possess character as well as being a delight to fly. An advertisement informing them that the little Currie Wot is for sale soon has them journeying 250 miles from home to see if it will meet their needs.

If their hearts had dreamt of a Tiger Moth or Stampe, unfortunately individual means did not allow either to realistically consider such classics, but in the Currie Wot, here was a delightful and affordable second best.

Closer inspection revealed that whilst she had not been pampered, only minor tidying would be required to bring her back to a condition her original builders would have approved of. The Wot had been in the ownership of a syndicate and

Robin supposed that no-one member had been responsible for the cleanliness of an aeroplane that an individual enthusiast such as himself would normally place as being as important as the actual flying.

They poked and prodded, peered under, over and inside before insisting on a few demonstration circuits to see for themselves that it could actually fly. That done, and once satisfied, a deposit was exchanged with handshakes all round and, once content, they set off on the five-hour drive home.

Robin was somewhat aggrieved that none of the syndicate was prepared to offer to even bring her at least part way to Devon, but ever determined the following weekend he returned to Booker to ferry the Wot home to Plymouth.

He spent his evenings reading any literature he could find on the Currie Wot and perusing the chart spread on the kitchen table to determine the best route for the non-radio machine. Being unfamiliar with the exact range and consumption it seemed to him prudent to make a stop for fuel as the journey would be towards the limit of its fuel tank and so he called Phil Cottle in Wiltshire requesting to put down on his farm strip near Devizes should it prove necessary.

Night after night he diligently followed the weather forecast and removed his prized old flying coat and helmet from their cellophane covers in the wardrobe.

On the day of collection, Robin's anxious anticipation reaches its peak when finally at the hangar, with all pleasantries complete and money handed over, he stands to watch the little red and white biplane wheeled out of the hangar almost ready for him to fly home.

His hope of a couple of familiarisation circuits of Booker before finally setting off is dashed by a stiff cross wind and both runways are in use by numerous gliders and tugs. It was a summer's weekend after all and *perhaps*, he thought, *it would be the lesser of two evils to just clear off and find out her idiosyncrasies on the* route home.

With a large wad of notes that had been the balance of purchase price bulging in his pocket the syndicate's leader swings the prop. The Currie Wot fires on the fourth attempt omitting a mist of oil smoke from one exhaust that eventually clears as the Walter Micron engine warms up. A short taxi to get the feel of her on the ground and a cockpit check, though there is very little more he can do than wiggle the stick and make sure the right bits move in the right order.

'HAPPY?' someone shouts.

Robin nods.

The fuel tank is full to the brim though whether it will be enough to reach Phil Cottle's farm strip remains to be seen. That is all part of the adventure after all. This is no Piper Cherokee; this is no ordinary weekend flight; this is a journey into the unknown.

Angled into wind for a run-up the full power check makes the little airframe tremble with the two sets of wings dancing slightly up and down. The tiny brakes on the motor scooter wheels hold well as he peers at the rev counter, though he has no handling notes to tell him what revs to expect anyway! So far so good: temperature and oil pressure look fine, but his heart sinks when trying each mag in turn – the switch over resulting in a very unhealthy mag drop. He queries one of the syndicate.

 'It always does that, but it'll clear when it's been flying for a bit,' yells the one with the money stuffed safely in his pocket. 'She'll be okay.'

He backs away from the cockpit and out of the slipstream with a grin and a thumbs up as if to say: 'Get on with it!'

Now it dawns on Robin that he is miles away from home in a strange aeroplane at a strange aerodrome and he begins to wonder if it will be so very clever to attempt to complete the journey.

No matter. There is ample room to abort take-off if necessary, especially in such a tiny aeroplane he can climb to cruising height and be within gliding distance of the aerodrome if needs be. With a final glance around the circuit he steadily opens the throttle and eases the tail up as she gathers speed. There is ample rudder to keep straight and she skips into the air upon reaching 40 mph after very little ground run. A steady climb to 1500 feet, before levelling off and adjusting the throttle for a cruising speed of 75 mph. Cautiously he tries each of the little mag switches in turn. This time, there is barely any variation and now a much happier pilot, heads off on a westerly course that will take him over unfamiliar territory at least to begin with. Peering down, he can make out the distinctive curve in the river that is Henley-on-Thames.

Realising he is a mile or two south of his intended track, he at least has a definite pinpoint and knows that even if he just heads vaguely south-west he will come upon the M4. The spring assisted trim is proving to be of little use however, and he soon gains an unintentional 500 feet whilst finding a comfortable position for the map on his right knee with two fingers preventing it from rushing off into the slip-stream and the remaining two grasping the top of the stick.

The cooling towers at Didcot become visible a few miles off to the right and at about the same time he catches sight of the M4 winding its way across southern England off the port wing. As he tracks toward Membury, he can begin to savour the experience of what he now knows to be a splendid machine!

Tiny though this one is, any biplane always gives the impression that there is so much more of it than a monoplane. The extra set of wings as well as the struts and wires make it seem much more substantial. Comfortable cruising speed seems about 75 mph but 80 mph requires so much more throttle opening to overcome the build up of drag that he decides to settle for the lower figure for now despite having read it as 80 mph in air tests.

His confidence growing, he begins to feel that this is an adventure far removed from the pioneers of the past, but in a minute way he can identify with them; empathise with the problems they must have encountered. And they didn't have to concern themselves with the congested air traffic lanes of the busy South east – lanes almost as congested as the M4 below.

Before long, the fear of the civilian air traffic lanes is superseded by that of the military zones over England's midwest; Wiltshire is below. Giant C-130 Hercules transports from Lyneham bearing down like juggernauts; Chinook choppers loaded with troops from Middle Wallop; or a Lynx carrying the top brass on a recce over the Salisbury Plain Training Area. *Good grief! Don't even think about it! Just watch and be mindful.* He kept going, overflying the disused airfields at Membury and Ramsbury and on toward Devizes.

His attention is diverted by a group of walkers trudging across one of the ancient chalk tracks of the Marlborough Downs. They stop to look up at the funny old-fashioned aeroplane humming on its way above them. Entering into the spirit of things he rocks the wings and waves down to them learning in the process that it is not particularly easy or sensible to stick your arm straight out into a 75 mph slipstream.

Again, he checks the mags despite a temptation to leave well alone and is pleased that they both seem to be functioning perfectly.

From Devizes, following the canal toward the famous Caen Locks and then in the distance the main lines of the railway until he spots the distinctively coloured barn that marks Philip Cottle's farm strip. Robin decides it is time, as arranged, to put down for fuel and a cup of tea. Philip, a keen pilot himself, operates a Cub and welcomes fellow flyers such as Robin with the opportunity to land at their own risk when necessary.

After one dummy approach to try to get some feel of the handling, he lines up again then notices Philip's car has pulled into the field. A first landing with this type and an audience too! Memories of hours on tail wheel Auster and Stampe several years ago pressure him to concentrate hard as the green narrow strip rises up to meet the scooter wheels. An aircraft such as this one could easily bounce the entire length of an aerodrome if not rounded out just right though he need not have worried. The little Currie Wot settles down politely and comes quickly to a stop as the tail skid drags through the grass. Gingerly, he taxies clear, fearing that at any moment he might tip her up on her nose as a result of relaxing too soon.

Switching off, he clambers out to greet Philip, aware that his grin must be ear to ear. Sheila Scott could not have been more elated when she finally landed in Australia! Though this first leg had lasted barely an hour, in his mind, he feels justified in meriting a hallowed place amongst such revered pioneers as Ms. Scott.

It is already late afternoon however, and after topping up with fuel and swigging a much needed cup of tea courtesy of Philip, he telephones Pat at Plymouth to report all is well so far, and gives him an estimated time of arrival.

His spirit however, loses some of its exuberance when it becomes apparent that the hot engine is not going to restart easily. Twenty minutes of determined swinging on Philip's part results in little more than a tired phutt-phutting and little else. There is a problem for sure that will have to wait for a later examination, but now with the evening closing in, it's important to leave Wiltshire behind and push on for home.

Just when it's beginning to look hopeless, the engine fires much to Philip's relief and not least Robin's. Once finally airborne into a glorious blue evening sky, Robin has little trouble picking up his next landmark – the TV mast stood high upon the Mendip hills.

Airborne, the engine seems fine and Robin relaxes with every passing moment of the experience. Again, there is time to relax and enjoy the superb visibility that reveals the Bristol Channel to his right and Glastonbury Tor just left of track and silhouetted by the lowering sun. Such is the God blessed artistry of the landscape below. He could sketch it but, oh, if he could write about what he saw. He had read every descriptive word written by Harald Penrose of his flights in this area in another Currie Wot biplane, *Airymouse*. Tonight he is all too well aware that he is flying (almost) in the shoes of Penrose and that is elation in itself. His only regret being that if only he had the ability to write about the scenery, its history and abundant wildlife as Penrose had done.

With Taunton behind him, he can cheat on map navigation and follow the curve of the M5 Motorway towards Exeter. His speed similar to that of the cars below he becomes aware of faces peering up from windscreens and sunroofs with occupants probably informing one another that Biggles is flying again this very night to save the empire. God save the Queen!

As the sun drops below the level of the top wing, a couple of minutes are spent cavorting up and down whilst feeling for sunglasses zipped up in one of the many pockets of his flying suit – courtesy of the RAF and his personal procurement officer, Squadron Leader Roger Wilkin. God bless Roger!

His groping is aggravated by the lack of elevator trim which is now getting annoying as the nose pitches up on relaxing pressure on the stick. With the sun dead ahead above the engine cowling he peers cautiously for aircraft and gliders spending a glorious summer's evening out of Dunkeswell, and so skirts north of Exeter unsure whether the airfield will be closed by this time.

As the engine is still something of an unknown quantity, it

seems wiser to fly south of Dartmoor rather than take the direct route over such a geologically unfriendly area. Picturesque though it may be, there are few places on its granite-covered slopes where a pilot can safely put down in the event of a problem.

Footnote

Later examination showed that worn piston rings were allowing oil down into the cylinder heads not an uncommon problem with this upside down design of engine but fortunately not impossible to rectify.

*G-APNT Airymouse at Dunkeswell, Devon, January 1987
(credit Wingspan Publishing)*

Inspired by famous books

Some time later, Robin actually came across *Airymouse* in a barn. I include the following extract written by Robin, which was originally published in *Popular Flying* magazine.

"I have been privileged to fly over much of the countryside

of which he has written, though unfortunately with attention directed to chart, compass and oil pressure gauge, and it has taken his writings to really open my eyes again.

"I became acquainted with Airymouse in April 1985. My colleague, Pat Crawford, and I had been on the lookout for a Curriewot to convert to a World War one SE5 replica to go with our Fokker Dr1 triplane. The Wot design had been used by several builders as a basis for a 7/8 scale replica of the SE5; strange how the wheel had turned full circle bearing in mind that it was John Currie's experience on such aircraft in the Royal Flying Corps that had inspired his Wot design in the thirties.

"We received a telephone call to say that a Curriewot in need of much care and attention was available in the North East of England, but when this turned out to be G-APNT Airymouse, with all its history it would have been sacrilege to turn her into an SE5 replica.

"Thoughts of flying the aircraft back to the Southwest were dispelled when we inspected her in a dim barn on a farm where she had stood for some time. Unfortunately, her previous owner, Les Richardson, had died the previous winter after having been a most active member of the Popular Flying Association in the Northeast.

"The aircraft log books revealed quite a chequered history over the years and we decided it would be prudent to de-rig and transport by road back to Devon. She had even suffered the indignity of a sunburst paint scheme; great for a Pitts perhaps, but now it was our intention to return her back to the original Hampshire Aero Club colour scheme worn in those earlier years. Fortunately, I was able to consult both Viv Bellamy and Harald Penrose over the many details and information needed and inevitably the tidy-up that was originally planned became almost a full restoration. Considerable help was given by Dave Silsbury whose help and advice has been sought by many builders of aircraft in the Southwest; the magnificent propeller now on the aircraft being one of Dave's 'specials'.

"Airymouse is now once again able to feel the air under her wings home again in the Southwest of England. Power is nowadays given by a continental PC 60, Ground Power Unit conversion of about 70 hp. This gives a rate of climb of around 600 ft per minute and a cruise of 75 mph.

"Basic aerobatics are a pleasure and of course lack of brakes and a tailskid remind one that this design is some fifty years old. Though the skies never did become full of private owners in their Curriewots, the design did fulfil the original requirement of the affordable aeroplane for enthusiasts; a quest that is still on today with designs such as the ARV Super 2, the modern equivalent of the Wot.

"Finally, many people are still curious to discover the origin of the name 'Wot'. Apparently, when the prototype was under construction, John Currie was constantly being asked, 'Wot you going to call it?' To which his reply was, 'Call it Wot you like.' And this was the name that stuck."

Currie Wot G-APNT "Airymouse" at Eastleigh, Southampton, 1958

Harald Penrose actually read the article and replied to Robin by letter:

Dear Robin Bowes,

I was of course delighted to see your story of the Wot, and Airymouse in particular, in Popular Flying. Sorry I missed seeing her at Henstridge in the spring, but hadn't realised there was a meeting. However, on several occasions I have looked up in hope of seeing the triplane! Where do you keep the couple?

A minor point which might add 1 mph to Airymouse(!) And slightly add to appearance, was that I faired the U/C round tubes with spruce (or balsa?) Tails encased in fabric. Interplane struts were painted white. Have been trying to find 3/4 view negative to print for you, but can't.

Regards,

Harald Penrose

Robin replied:

Dear Harald Penrose,

Many thanks for your letter passed on to me by the P.F.A., I am sorry that since my original telephone call to you almost 2 years ago I have not come up with any progress reports but I have only just got Airymouse back in the air for this season.

Since the photos were taken that were used in the P.F.A. magazine I have in fact faired in the u/c legs which has improved appearance but no measurable change in performance! I have also added "Hampshire Aeroplane Club" to top of the engine cowlings, something that is there in all the very early (1958) pictures of the aircraft, but I don't know if this remained during your ownership?

Obviously there must now be a number of changes since you operated the aircraft but I have tried to keep the colour scheme and appearance at least in the original spirit of Airymouse. You would I am sure be delighted at the reaction from many people who have come over to the aircraft and asked "Is this really Airymouse?" Like me they have been enchanted by the book and the way it captures their feelings and thoughts whilst flying for fun, my regret is that I don't seem to be able to put it down on paper!

I took the aircraft to the P.F.A. Rally at Cranfield on 3, 4, 5th July and again it was a pleasure to listen to people's reaction to seeing her. I also met Lynn Williams who did the cover painting for the book. I have been asked to take [Airymouse] to Badmington Air Day on 26th July, where I shall also be displaying the Fokker triplane, as they are planning a fly-by of several significant P.F.A. types. At present Airymouse lives at Dunkeswell aerodrome, though I do intend to bring her nearer my home eventually, though suitable bases for non-radio, brakeless aeroplanes are rather limited these days! I would love to arrange for you to be re-united with her again if you wish. I don't know what reaction we would get from Yeovil these days, but I would happily bring her over there or any other venue that you may suggest.

Meanwhile many thanks for taking the trouble to write to me, and I hope that the three of us can get together sometime!

Best wishes

Robin Bowes

Chapter 7 – Doing Business in Ermington

November the 25th 1977

A colleague of Wendy's at Westward Television in Plymouth was a sound engineer and together with his wife they ran their home as a guesthouse. With the oncoming problems of age they felt that the demands of running a popular guesthouse was becoming too great; and so they offered the Old Inn house, Ermington, for sale to Wendy and Robin.

For sale that is with the proviso that Wendy and Robin continue the guesthouse business that the couple had so lovingly developed over many years. Historically, it was a purpose built inn and it wasn't difficult to imagine an earlier pre-motorised age where weary travellers arriving in the centre of the village en route for, or departing from, nearby Plymouth would book in to rest awhile.

Although Robin wasn't looking to become a guesthouse keeper, the Old Inn house was very attractive in that not only was it a spacious, historic cottage, full of character and a world away from their modern semi in Plympton – it had off road parking for his cars under the archway and a very large, hangar like garage – large enough to be a proper workshop and accommodate a multitude of cars or even the odd aeroplane. It was just a shame that for the second time he'd had to sell his beloved E-type in order to raise his share to purchase the Old Inn house as an E-type would have fitted in here very well indeed.

For her part, Wendy had decided that she would resign her post with Westward and concentrate full time on the guesthouse.

Westward were not keen to let her go so easily and

generously offered her a six month sabbatical just in case her new role of being a hostess didn't work out, but Wendy was determined that if the guesthouse was to be run properly, then she had to cut her ties with her old life.

And despite the expense in investment, there was also no regular income. Robin was selling cars, but not every week was a good week; the second-hand car trade being notoriously up and down in its fortunes. The guesthouse would have to be their main income and at least that first summer of 1978 they were expecting the regular clientele that had been coming to stay over many years.

Returning to work at Westward for three days a week, Wendy split her time as a bed and breakfast host until such time as the work became too much, despite help from the previous owners. And so ended the first and only season for the fragile collaborative effort. Wendy closed the business to guests and returned to Westward and her old post working full time. The Old Inn house was now most independently Robin and Wendy's to do with as they pleased.

Robin had actually enjoyed his brief time as a "wine waiter" and had sold bottles that he recommended most highly to those who would otherwise have classed themselves as tea total. He was a charming host and when not at work in the garage or racing the latest production Chevette HS, he would be at various venues playing drums either in the Russ Thomas band or Plymouth Show band. But it was the large garage that took up the biggest part of his time. He loved it: painting the concrete floor a brick red, marking out the parking bays and painting the walls white. His friends had joked that you could safely eat your lunch off the floor! Jokes apart, the garage was his pride and joy.

In those days, a visitor to the Old Inn house might have wondered whether there was ever anyone in. Wendy worked long hours at the studios in Plymouth, with the bulk of broadcasting taking up evenings. Westward produced ITV regional programmes for the Southwest including for children's hour: *Gus Honeybun's Birthdays*; then at 6pm there was *Westward Diary* the regional news magazine programme hosted by avuncular Ken Macleod, supported by a large cast of very talented presenters including a young Angela Rippon in the years before she found national fame as a BBC newsreader.

Regional television in those days, especially in the case of Westward, could produce some of the most entertaining programmes of the evening, and *Westward Diary* as a magazine news programme was a case in point. Presenters like Macleod, broad shouldered, mature and immaculate in appearance, came across as jovial bon viveurs. He was a larger than life host with a tremendous talent for the delivery of lines with natural comic timing that could have the studio in hysterics during transmission.

On those evenings, Robin might be found, depending on the season, either at work in the shed, flying or playing in a band. The shed was the most likely place especially during the winter months; it had everything he wanted – space – being more than large enough to accommodate some of the most professional tools and equipment on the market. And sometimes to Wendy's annoyance, he would avoid visitors ensconced as he might be in whatever project was underway. He had little time for cosy afternoon tea parties and the trivia that went with them. Sitting around drinking cups of tea and passing the time of day even with old acquaintances was not for Robin and so Wendy found herself making excuses for her errant husband.

She even began to make excuses to herself, as it wasn't only the trivial occasions that Robin was avoiding. Their respective life paths were not converging and they saw little of one another as each in turn made excuses not to attend certain events. If Robin wasn't in the shed, where at least she could find him, he was flying or racing or playing in the band. All these things that he did had initially attracted her; he was such a capable man and she loved that part of him and deeply admired his abilities, but these same qualities were now pulling them apart.

When he brought Wendy home from hospital following an operation, he'd gone to great lengths to decorate the hallway with "Welcome Home!" Inscribed on sheets of toilet paper draped from beams and various walls. She appreciated the gesture at first that is until he told her that he had to go and remove an aeroplane from a field in which he'd had to make an emergency landing.

'I'm sorry, but I've got to go out. I've been flying and the engine stopped on me and I had to put it down quickly and on landing I've broken the wing and I've knocked the nose wheel off. We hit a molehill and that was enough to tip it over.'

The aircraft in question was a Tipsy Nipper – a small, mid-wing single-seat aeroplane with a Volkswagen engine. The previous owner had done some work on the carburettor, but used gasket cement instead of a paper gasket when replacing the carb. Minute particles of gasket cement had come lose and blocked a jet so stopping the engine in mid flight. The incident did nothing to ease Wendy's fears when Robin was flying and two days later her stitching burst leaving her very poorly indeed.

The Nipper was repaired and quickly sold on, but there would be others to follow. An Isaac's Fury was the next acquisition and when they needed information on the Fury, Robin and Pat contacted its builder, John Issacs, who was based at Land's End Aerodrome. They agreed to go and meet him and it was on this trip that they spotted the Fokker triplane standing redundant and covered in dust in one of the hangars.

Robin himself was not immune to the odd injury through twists and strains of one sort or another and one night working in the garage he found himself barely able to walk and very reliant on Wendy. He'd injured the cartilage in his knee and in order to get help he had to hobble with great difficulty across the courtyard to the house.

To Wendy's horror, Robin staggered into the kitchen with tears pouring down his face. He tried to reassure her.

'Don't make a fuss! I'm fine.'

'You're obviously not fine, you can hardly walk.'

'Wendy, don't make a fuss!'

'But you've got tears running down your face.'

She made him as comfortable as possible before ringing her father, a chemist who advised they seek medical advice as soon as possible. Within 24 hours Robin's knee was being operated on.

During the winter months, to make sure he was as warm as possible when working in the cavernous garage, Wendy invested in a pair of Damart long johns for him, which he initially rejected.

'I can't wear those!'

'Why not? Nobody will notice and they'll stop you getting problems with your joints and your back. I won't tell anyone if you don't.'

He later admitted that the combinations were the best present she'd given him.

The Old Inn House, Ermington, Devon

Instructing

There were times when things reached a point where they could see no future in the Old Inn house and so they put it on the market, not that Robin wanted to lose the garage; it was ideal for his car sales, but sales could be sporadic and so Wendy suggested he might like to try other avenues such as becoming a flying instructor. On the surface, this seemed a good idea.

'You've got to decide what you want to do,' she told him. 'You've either got to put your time to the car sales, or if you want to go flying, you've got to think about doing it properly – seriously, as a living.'

To qualify as an instructor would mean a week spent in Jersey on a training course. It would be a nice little break in itself, not that Wendy could go with him.

On the morning Robin left Ermington for his flight from Exeter airport, Wendy was sewing some trousers when there was a knock at the front door. She put down her sewing and went to open it, but couldn't – it was jammed solid. On the other side, Dan Davis the village postmaster was trying to inform her of the calamity that was unfolding in the Square.

'Wendy? It's me, Dan – from the post office. There's a lorry stuck in your doorway and I think he's done some damage. Turn the handle your side and I'll push.'

The articulated lorry had embedded the rear end of its trailer into the front door of the Old Inn house and now the door wouldn't open with one of the door panels well and truly smashed.

Trying desperately not to panic Wendy rang Exeter Airport.

Fortunately, Robin hadn't taken off.

'Oh, Robin, come home will you? The back end of this lorry's embedded itself in our door and all the stuff's come away.'

'Well I can't, I'm about to take off for Jersey.'

Despite Wendy's pleas, Robin refused to come home saying there was little he could do in the circumstances and a builder could put it right quickly and under the home insurance policy.

Although Wendy initially felt abandoned, she later realised that her husband – though notoriously stubborn – was at least confident that she could cope perfectly well without him; and on occasions like this, she did.

The week's course in Jersey didn't result in Robin becoming a flying instructor, but it did concentrate his mind on the purpose ahead. On his return he told Wendy:

'If I've got to be a flying instructor everyday for a living, I don't think I want it. I enjoy doing it, but if I have to do it everyday I don't know if I'll enjoy it so much.'

Aerial pictures

To Wendy, the approaching sound of the triplane overhead sounded like a tractor in the sky. Stepping out into the courtyard she'd look up to see Robin in the cockpit waving. She suspected he waved not only to her but to neighbours who were also stepping out to look up at the strange sight of the Red Baron flying victoriously over their small, Devon community. On fine, clear days he would take aerial pictures of the village and pass on the prints to those who were interested.

And in turn, people were taking pictures of Robin, particularly the media. Robin's aeroplanes were making the news programmes on both ITV and BBC regional channels. "Airymouse" and the Fokker triplane with their respective histories were ideal subjects for television news programmes. In aviation circles replicas like the triplane were viewed as a wondrous thing to behold and as its popularity spread the RAF were even paying Robin to attend their shows with it – a very rare honour for a civilian pilot indeed.

The BBC called while Robin and Pat were in the process of re-covering the middle wing in fabric while the "tripe" was on display in a hangar at RAF St Athan's museum. The BBC needed the triplane to feature in a new programme they were making entitled: *Reach for the Skies*. Robin and Pat were elated; it was just the sort of opportunity they'd been hoping for since buying the triplane from Lands End, but in respect of the media their negotiation skills were sadly lacking and, perhaps in awe of the mighty corporation, they practically gave their services away for a paltry £750 for what they believed would be six minutes flying. Filming was to take place in Fair Oaks, Kent, within 14 days.

In reality, the flying required by the director was far more than a mere six minutes. The flight to the location required was 50 minutes followed by half an hour flying at just below tree top height while being filmed from a helicopter and then a 40-minute flight to Fair Oaks.

At Fair Oaks, Robin was requested by the director to extend

his landing run, but with the triplane having a tail skid rather than a tail wheel, an extended landing could not be guaranteed with the flight controller warning him that they didn't have a grass strip as such, but to use whatever he could find with the understanding that it could be boggy.

Sure enough, on extending the landing, the triplane ran into boggy ground, tipped on its nose and promptly snapped the wooden prop. Various airport vehicles were dispatched to pull the triplane clear of the mud.

The airfield was closed as initially a Land Rover was dispatched, but that became stuck even before it reached the triplane. Then a tractor was sent out and that too became stuck. The fire tender was sent across as the triplane was seen to be leaking fuel and a big American built heavy lifting truck with a hauser was dispatched to winch out the stranded vehicles. A comedy of errors then ensued as the recovery truck pulled valiantly to release the Land Rover without the tow cable being attached to the stricken vehicle. The cable was rewound after considerable effort, but even when the connection was eventually made, the Land Rover driver in an attempt to assist his colleague, accidentally selected a forward gear instead of reverse. Pulling in the opposite direction he duly succeeded in churning the mud into an ever greater quagmire as the vehicle became even more well and truly dug in; and all this under the watchful gaze of the CAA who were inspecting the airfield that day.

Pat and Robin succeeded where the rescue vehicles failed using little more than a stepladder and a rope they pulled the triplane by the tail manhandling it backwards out of the mud.

Even out of the mud, the triplane wasn't going anywhere with a broken prop, so they left it at St Athan for two weeks while a new one was made. The BBC was quick to honour their side of the contract paying as agreed £750. The repair to the prop cost £780 in addition to the hangar fees for two weeks.

On the way up from Devon, it had cost Bowes-Crawford Flight two and a half hours flying time to Fair Oaks. In addition, another five hours for filming and the return home coupled with a crippling £50 per flying hour for comprehensive insurance as the triplane had no brakes. It was a loss of colossal proportions. National exposure was guaranteed by the transmission of the programme, but at great financial cost for the partnership.

Such was the lot of many of the civilian display pilots flying classic aircraft.

'I only do this for the fuel,' a Spitfire pilot told Pat. 'I earn £30,500 a year, and the insurance is £30,500. So as long as they top me up with fuel that does me from one show to the next, and I've still got my Spitfire.'

The first show with the triplane

The two-day fighter show at North Weald was to be the first outing for the triplane as a flying display aeroplane. Robin was to fly up whilst Pat drove his own car loaded with back-up equipment including a jockey wheel, the plan being to rendezvous at Popham airfield in Hampshire.

En-route, Robin was to be diverted in a way that was to become quite common whenever he flew the triplane across country. On identifying himself to Exeter Air Traffic Control as 'Golf – Bravo Echo Foxtrot Romeo – Fokker triplane,' he was requested to do a flypast.

'Say again your aircraft, please?'

'Roger. Fokker triplane.'

'Please divert to fly past the tower would you?'

In a pattern that was to be repeated all across the UK and

Europe in the years to come, ATC would hold their traffic while Robin did a low level flypast just so that the controllers could press their noses to the glass to gape at an aeroplane that they must have thought extinct and consigned to history. But it was true; the Red Baron was flying again.

In accordance with Pat's recommendation, Robin was running the radial engine in at between 1750 to 1800 RPM, but it was causing the plugs to oil up resulting in a misfire so he had to put down at Henstridge airfield.

On the ground, he set to work removing the bottom two plugs allowing the oil to drain down past the piston. If this wasn't done, a hydraulic block could occur when the engine was turned over manually and be serious enough to blow the barrels off and do irreparable damage. As he was wiping the plugs dry he became aware that he'd attracted an audience – two young boys were watching him.

'Do you know if anyone's around? Any mechanics?' He asked them.

'This isn't an airfield anymore mister, it belongs to the BBC – they bought it.'

No matter, Robin was a very able mechanic. However, when he pressed the button to fire up the engine, the battery was flat, so, resplendent in his dashing WW1 flying gear, he left the triplane and walked out onto the lane that passed the airfield with the intention of flagging down a passing motorist. He didn't have to wait long; an elderly couple, initially unsure of just who was waving them down, stopped out of sheer curiosity to speak to a man dressed as the Red Baron.

'I'm sorry to stop you,' announced Robin, 'but my battery is flat and I need a jump across. I've got some leads, would you mind?'

'Of course not, young man. We'll do that for you,' replied the driver.

The driver's exact expression was not recorded for posterity, but it's safe to presume that he and his wife expected to be taken to another car in distress and not a scarlet Fokker triplane parked outside a hangar at an airfield; and there can't be many drivers who can claim to have assisted in jump starting a Fokker triplane, but such was the state of the world in the days when Robin Bowes was flying as the Red Baron.

The jump-start did the trick and the Warner Scarab fired into life causing the triplane to begin to trundle off gently. With barely a moment to shout thank you and wave an acknowledgment in gratitude to the driver and his wife, Robin leapt onboard and promptly strapped himself in before opening the throttle for take off. It is not known whether the couple that came to his aid were ever taken seriously when recounting their strange story to friends and relatives.

The Popham Rendezvous

When Pat arrived at Popham the day was already becoming quite hot. He was hoping he'd be ahead of Robin as he needed to catch the triplane's wing on landing – the "tripe" having no other way of braking and the grass strip at Popham had a notorious dip in the middle like a worn old sagging mattress meaning that without brakes it could pick up speed running into the dip.

He was met by Jack the café owner who told him that Robin had reported some problem and had put down at Henstridge, but with the problem fixed he was on his way again and should be within sight shortly.

Sure enough, some twenty minutes later the distinctive tone of the triplane's old Warner Scarab engine could be heard approaching. Pat asked Jack to run him out onto the grass field in order to catch the wing; but as Robin touched down Pat found himself unable to run fast enough to catch the lower wing and so the tripe motored on before eventually rolling to a stop opposite Jack's café whereupon it turned 90 degrees and rolled directly at the café wall. Somebody made a vain attempt to grab the wing, but it couldn't be stopped. Jack's wife rushed out to see a full scale red Fokker triplane coming straight at her. She screamed, turned and ran back into the café just as the triplane's prop made contact with a cork life jacket originally issued by the RFC and hung by Jack as retro décor on the wall. Although in a spin, had she remained rooted to the spot she would have been in a very serious predicament.

It took a good, strong cup of tea served in her very own café before shattered nerves could even begin to settle. Thankfully, the only skin broken was to be found on the prop of the triplane, so Jack generously put them up for the night and together with Robin and Pat he worked alongside them to midnight helping to replace the damaged skin on the prop.

It was an inauspicious start to the first outing. Wendy was far from happy when he'd left Roborough and now on the dawn of the first show, the clear weather had given way to mist and drizzle.

Fellow pilot and race driver, Ron Lambton, who was intrigued by the sight of the triplane, approached them. They told him they were going to fly at North Weald's fighter event but that conditions were going to be a problem.

'Do you know the way?' he asked.

'Not in this,' Robin replied.

'I'm going that way, you can follow me.'

So Robin took off behind Lambton's Super Cub and followed him up to the M25 where above Potter's Bar he was then able to talk directly to North Weald.

'Orbit where you are and we'll call you in,' advised the Air Traffic controller.

Robin's string of bad luck seemed to be behind him. The

Warner Scarab seemed fine, but the view below of built up suburbs didn't exactly put Robin's mind at rest. If there were to be an engine failure now over Potter's Bar making an emergency landing would be very tricky indeed. He hoped North Weald wouldn't keep him waiting too long.

Just as things seemed to be ticking over without problem, the Perspex aero windscreen ripped away from its mounting screws on the upper surface of the fuselage and blew past Robin's head like a missile. If it had smashed into his face it could have blinded him or knocked him clean out. His fear now was that in such a heavily populated area it could land on somebody or something causing considerable injury or damage.

Even though the windscreen is extremely small it's an effective air deflector and without it the wind was now blasting directly into Robin's face at over 100mph. With his free hand he pulled his goggles down over his eyes called up North Weald and told them of his predicament.

'If you've got trouble come in and land immediately,' they advised him.

By the time Pat gained entry to the airfield and located the parked triplane Robin was nowhere to be seen. A voice over the tannoy was calling for people to come forward with any Perspex they might have, as a pilot needed it urgently.

Realising that there might be some sort of connection with the announcement and Robin's absence, Pat enquired as to what was going on.

'He's had an accident!' came the response.

'What happened?'

'His windscreen blew off somewhere not far from here.'

'Has it cut his face?'

'No, but he's out looking for Perspex.'

A policewoman approached with some bullet proof Perspex of an inch thick – ideal for a dogfight; but it was three dirty faced boys who had the most intriguing offer – a Perspex bust used for displaying bras in shops.

'We can't use that!' Pat told them. The boys, laughing uproariously, were sure it would be a great joke for the pilot even if it wasn't in keeping with a World War One replica.

'Yes alright, well I'm sure we can use something.'

Robin had by that time found a suitable piece and also located Pat. They invited the boys to come and see the triplane and show them what was needed.

The boys were over the moon at the invitation to come into the special area reserved for the aeroplanes and their pilots. They'd never seen anything quite like the triplane and were now instant converts. In the years to come they would follow Robin's triplane to all the London shows they could get to.

Old Warden

Robin's first show at Old Warden, Bedfordshire, was not without incident either. Dipping the triplane's nose into a dive before executing a loop, the old Warner Scarab engine stopped with a bang. The crankshaft seized causing the propeller to split from tip to tip along all seven of its laminations like a matchstick that had been struck by a hammer. Robin quickly had to get the triplane down as safely as possible whilst the crowd below watched in awe the drama happening above their heads. The triplane is no glider and the urgency with which a safe landing can be made is as dependent on good fortune in respect of the aeroplane's position in alignment to the runway as it is to pure skill. Robin struggled to keep control, as the triplane was a difficult machine to fly with a working engine let alone a seized one that was now a dead weight in its nose.

Had the engine stopped at a different point in the display circuit, it would have been a very different story. As it was, there was much relief that Robin's skilful handling of the situation enabled a safe touch down. Safely back down on terra firma Robin and Pat pulled the engine to pieces to find the cause of the sudden seizure. Careful investigation revealed the culprit as number four cylinder that had caused the problem and so the engine was replaced with another Warner Scarab.

Even at Old Warden aerodrome, which is famous for its remarkable and eclectic collection of veteran and vintage aeroplanes, the Red Baron Fokker triplane always caused a stir being a remarkable sight for aviation enthusiasts. But not everyone appreciated the triplane.

Some days later following its engine replacement, Robin flew the triplane on to nearby Cranfield aerodrome but had to put down in a nearby field just short of the airfield. That evening in the bar someone asked Pat:

'Did you see that Mercedes turned upside down in the ditch?'

'Yes, I did,' replied Pat. 'Just down the road from here.'

'Well, whoever flew the triplane in caused that. The driver looked up and saw this Red Baron triplane, lost control and left the road!'

At Leicester, display pilot and Barnstormers Flying Circus founder member, Barry Tempest considered Robin's triplane display "too low and too slow," though he admitted that the Leicester crowd had "loved the performance with its element of danger."

The American airbases were incredibly generous when it came to providing food plus fuel and made sure that visiting pilots, their crews and aeroplanes were more than sufficiently catered for. Both commodities could be administered with a generosity rarely found elsewhere.

Sometimes though, it was all too much.

Robin only ever did aerobatics with the triplane carrying 7 or 8 gallons rather than filling to its capacity 25 gallons, so when asked by an anxious USAF tanker driver to take a staggering 800 gallons there was little that could be done other than to produce a couple of empty jerry cans from Pat's car and accept filling the tank of the triplane. It would mean a flatter show without the aerobatics, but it was a generous gift from the United States Air Force.

The driver explained how the Flying Fortress had been allocated 1200 gallons but was now full to capacity so there was 800 gallons going spare and it had to be delivered to anyone who'd take it, as it couldn't be returned to the ground tanks.

'Do any of your friends need fuel?' asked the driver.

Of course, everyone did.

Neurosis

The Americans though generous were also quite neurotic when it came to security. Each USAF base was a piece of US territory in all but name with currency, commerce, transport and policing all handled within the confines of the base. Visitors needed dollars for cash when making a purchase at the PX. Breaches of security were handled by American military police; children of service personnel were bussed to school in American built yellow school buses operating to American traffic laws. RAF Wetherfield was no exception: a US borough in microcosm, it looked after its guests most carefully.

The hospitality afforded to Robin and the triplane was a case in point. No sooner had he landed than the triplane was whisked away to a hangar whilst Robin was escorted to his quarters. There was no time for him to take out the bottom two spark plugs and drain the oil past the piston. Without this procedure on a radial engine, oil can foul the plugs and do irreparable damage to the barrels.

By the time Pat arrived and had been checked through security, the triplane was under lock and key with no access permissible until the following morning.

Following a very comfortable night and sumptuous breakfast the pair were collected from the officers mess and driven across the airfield to the hangar. Getting out of the car, Pat thanked his escorts and introduced himself and Robin to the ground crew who'd been watching over the triplane. He was about to enter the side door of the hangar when an authoritative voice stopped him in his tracks.

'Stand there sir. Don't go over that yellow line.'

Under orders to follow the ground crew carefully, Pat and Robin watched in amazement as they were lead into a bunker. Following a solemn but well rehearsed routine, a button was pressed that triggered alarm bells and flashing lights much to the concern of Pat and Robin. They watched in awe as the largest and heaviest main doors they had ever seen slowly opened to let in natural light; and there, sat alone at the back of the vast hangar was the triplane.

'Well, there it is. That's atomic proof!' joked Robin.

Pat could barely believe his eyes. Turning to Robin, he said quietly: 'If there'd been an atomic strike on this country last night, just think in 200 years from now somebody might have opened this hangar looked at that archaic, forlorn machine, pondered it for awhile then exclaimed, "no wonder they bloody lost! It hasn't even got real guns!"'

'This hangar points directly at Moscow, sir,' one of the ground crew explained.

'In the event of war, the F4 is unleashed. You'll notice the aircraft is chained, but there is no taxiway. The pilot fires up the engines, and he goes straight without turning at anytime on the ground or in the air.'

'And what would you fellas do?' asked Pat.

'We, the ground crew, are expendable. Once our aeroplane is gone, we've got nowhere to go. We live in the hangar which is all air conditioned.'

On the day of the Air Show at Weatherley, American generosity was again at the fore with an all day barbecue for pilots and their crews, the barbecue remaining open until the last pilot had gone home.

1987

For the third winter running, the triplane was placed in the museum at RAF St. Athan as a static exhibit, part of their Historic Aircraft Collection. Thankfully, no major work had been necessary and so Robin could simply dust off the wings and fuselage before a thorough test flight south to Dunkeswell where he would be in time for the Popular Flying Association's Fly-In on Sunday 10th May.

It had seemed a long winter lay-off and he was keen to rehearse the display routine again. May in the Southwest can be a good month, and this year was no exception with warm days and sunny skies, so there was every opportunity to beat up the circuit. He could imagine large crowds below craning their necks and licking at ice cream as above their heads the daring young man in his scarlet flying machine flew to thrill. He could almost hear the imaginary commentator reading from the carefully prepared notes.

Even as Robin was about to begin his third season the triplane reminded him of just how difficult a 1917 design can be to fly. It took time to reacquaint with its most awkward characteristics, but this was why he loved it. The triplane continually presented a challenge in the air and on the ground and it wasn't only the machine that took some getting used to.

At the beginning of each new flying season, he would feel quite ill for the first few times performing aerobatic rolls. It didn't matter what the aeroplane was, the ill effect was always the same; then it would clear and he'd be fine. This determination of his to break through the "sick" barrier every year said much about his determination.

The first minor mechanical hitch of the year was the tailskid, which promptly broke on landing. However, this was the best time for it to happen – during rehearsal. A quick call to Dave Silsbury soon rectified the situation. Dave, as always, was a rock and presented a stout ash skid – varnished, ready to fit – that he copied from the broken original, and all done overnight.

The first gig of the season on the 19th of May was for the independent local radio station, Plymouth Sound, which was celebrating a birthday and wanted to put on an air display for staff, guests and listeners. For a fee of £350 Robin as the Red Baron would make his "entrance" at 8 pm over the Royal Western Yacht Club just as the Royal Marine Band were finishing their marching display. Of course, such things as a flying display over Plymouth water took some arranging. Robin had to contact the Civil Aviation Authority to officially notify them of what he intended to do by way of "unusual air activity" – incorporating aerobatics and the flight line that he would be taking. He would also need to notify Plymouth Airport as they were the nearest Air Traffic Control, and Devon and Cornwall Police also had to be notified – though Plymouth Sound could do that.

On the night, Robin, as usual, did more than simply turn up and fly a few loops. Without prompting, another aircraft towed a "Happy Birthday!" banner and, most remarkably, a Nuclear Submarine arrived in the Sound as if arranged by Robin himself. The station's managing director was most impressed referring to Robin as a "magnificent man in a flying machine!"

The Red Baron

Robin meets von Richthofen
1989-1993

It was perhaps inevitable that flying as the Red Baron Robin would invariably attract the attentions of those who were fascinated by the legend of the famous fighter ace.

At displays, Robin was always happy to stand by the triplane when it wasn't flying and talk to those spectators who were fascinated with it. The questions were often predictable especially after a few seasons on the air show circuit.

'Does it fly?'

'Are those real machine guns?'

'Is this the same aeroplane that the Red Baron used to fly?'

'How fast is it?'

'Where was it made?'

If Robin tired of the repetitive nature of some of the questions, he never let it show, nor did he hide away in the pilot's enclosure and let a notice board do the talking for him. For Robin, the spectators were the important element of any event as without them there would be no show and he valued their interest most highly.

Occasionally, there would be a fan who had more than a passing interest in the triplane and the legend of the famous fighter ace. One such fan asked Robin and Pat to pose in front of the triplane as he snapped away photographing every aspect of the machine including the cockpit and the engine.

'Do you know what von Richthofen's dog was called?' he asked Robin.

This was a new one! In all his extensive research, Robin had given rather less interest to the Baron's dog and constant companion, though he was confident it wouldn't have been called "Snoopy".

'Sorry, I don't know,' replied Robin.

'He was a Great Dane and his name was Moritz. The baron adored him.'

Any comparisons between Robin's flying career with the triplane and that of von Richthofen was never likely to include the ownership of pets or mascots, as Robin had no interest, but he wouldn't forget the name of Moritz again should the question arise a second time.

'You know that von Richthofen was awarded the Ordre Pour le Merite?' asked the fan.

'Yes.'

'I have the replica,' replied the fan. 'The original wasn't worn normally because of its value, so pilots wore a replica. I have von Richthofen's replica. It cost me a lot of money, but it's worth £20,000 now.'

Any remnant of von Richthofen's short life was much sought after by collectors and the fashion of collecting anything to do with the Baron had begun on the very day of his fatal crash in April 1918 when his wrecked triplane was pulled apart for souvenirs.

At one base just before an afternoon show with the triplane, Robin and Pat went for an early lunch in the officer's mess. The mess was empty with the exception of an RAF pilot sat alone at a table.

'Do you mind if we join you?' asked Robin.

'No, I'd be glad of the company,' replied the pilot. 'Are you flying later?'

Robin told him he'd be flying the triplane shortly.

'Really?' said the pilot before fumbling in his uniform pocket. He then placed on the table a small piece of scarlet fabric. 'What do you think of this?'

'Is that what I think it is?' asked Pat.

'It's from von Richthofen's triplane taken when he crashed and it's authenticated.'

Robin and Pat examined it in astonishment.

The pilot continued: 'My father gave me this. I've always kept it in my pocket, but I never thought I'd ever see a Fokker triplane. You're the only people I've ever shown this to.'

Such is Legend

Pat was at the Old Inn House, Ermington, when there was a knock at the door. From the kitchen, he could hear Robin welcoming someone followed by profuse apologies from an old gentleman who was enquiring as to whether Robin flew the triplane.

At first, Robin thought he was going to be admonished for low flying, but the old man's visit was of a very different nature. He explained how his father had been the official photographer at von Richthofen's funeral and that he had pictures that might be of interest to Robin. It was fortunate that Pat was there as it also gave him an opportunity to see these remarkable images.

The old man described how von Richthofen's original grave had been dug in the corner of a French cemetery at Bertangles much to the annoyance of local people and despite all the ceremony of a full military funeral with a rifle salute from Australian soldiers.

'It was just a rough old hole,' the old man told them, 'and they buried him and erected a rough old cross made from a propeller and chucked all these old stones on top of bits of slate before being bedecked with wreaths.

'When the war ended shortly after, the Germans wanted all their heroes returned and so the body, or what was left of it, was exhumed; but somebody had got there first – probably angry locals. There was a local man who used to be mayor and people thought he was crazy. He told the Germans that von Richthofen's body was not in the churchyard, and when they exhumed the coffin, only part of the body was there – his head and shoulders. So they took what was left and he was buried twice more – first at Fricourt before being moved again to a final resting place in Berlin.'

Throughout the twentieth century – and it's probably still true today – remnants of Baron Manfred Freiherr von Richthofen's life continue to turn up in the oddest of places, like the flying boot that was donated by a woman to an Australian museum to match the one already on exhibition. She explained to staff that her brother had kept it during his lifetime but now that he was gone she would donate it the museum. Such was the legend of the Red Baron.

Von Richthofen's grave at Sailly le Sec, The Somme

A Static Arrangement

For convenience of travelling, the triplane was now housed for the winter months at the Fleet Air Arm Museum at Yeovilton, Somerset, where it was on display as a static attraction. This could be a good arrangement as large numbers of visitors saw the triplane; but it was never easy to extract the triplane for a flight as it involved gate passes and a whole team of naval ground crew who would escort Robin and Pat from the main security gate.

Once signed into the base, it could take anything up to two hours to get airborne as the fire section had to be informed and other static displays would have to be moved to extract the triplane. The fire fighters would then guide Robin out to the only grass section where he could take off and land with a tailskid. In all, it was a very protracted affair for a short flight. It would then take another two hours after landing to put it away again.

The Invitation

One advantage of the triplane being at Yeovilton was that it would be an ideal place at which to formally invite the German Ambassador to Great Britain, Baron Doctor Hermann von Richtfhofen.

Freiherr Dr. Hermann von Richthofen was taking his own place in Anglo-German history in respect of being the very first ambassador to Great Britain of a newly united Germany. The Cold War was coming to its close and the Berlin Wall was being dismantled. London had accommodated two German embassies one for the German Democratic Republic under Erich Hoenicker and the other for the Federal Democratic Republic (West Germany).

The London Evening Standard once dubbed the ambassador as the "Cindy Crawford of the Diplomatic Corps." He was a roving ambassador in the true sense of the word and liked nothing more than to be out and about representing Germany in a country that was still to shake off entrenched attitudes that had taken two world conflicts to mature.

Robin had written to the German Embassy inviting the ambassador to see the triplane at Yeovilton should his busy schedule ever allow. The invitation was passed on to the ambassador who wrote back thanking Robin for his kind invitation, but expressing concern that the press might in some way misrepresent such a visit.

Ever determined, Robin was sure that such a visit could be kept low key and reassured the ambassador who was happy to accept providing it remain low profile.

The Ambassador did not want a visit to be construed as a re-enactment of some past national glory based on the wartime actions of his famous great uncle. In his visit, he was simply acknowledging a rare phenomenon – a British interest in the German cultural archive; and Robin's triplane was playing a significant part in that interest.

Robin was both delighted and honoured that the

ambassador had accepted his invitation. It would, after all, be a great opportunity for Anglo-German relations – relations that had historically been good right from the awkward beginnings of Deutschland as a unified confederation of states up to the outbreak of war in 1914.

In pre-confederation days of the nineteenth century, Prussia was a loyal ally of Great Britain and Queen Victoria herself was the most famous face of that unity. Her Hanovarian ancestors had been in part invited to the British throne to resurrect that lost Saxon line that had died out with Edward the Confessor and been trampled into the English fields at Pevensey in 1066.

Even during the "Kaiser's War" of 1914-18, the fighting men in the trenches on both sides – "Tommies" and "Huns" – had questioned just what sort of conflict was ripping apart the ancient fabric of blood bonds that tied Anglo-Saxons and Celts to their continental cousins; it just didn't seem right for such former allies to be engrossed in mutual destruction.

Come the day, what Robin had not expected was the entourage that accompanied the ambassador coupled with the intense media interest in the visit with both TV and press coverage taking more than a casual interest. The media had no interest in Robin and were practically pushing him aside to get to the ambassador who, sat in the cockpit of the triplane, was providing an excellent photo opportunity while fielding some of the most banal questions:

'What would your great uncle say if he saw you sat in a red Fokker triplane, Mr. Ambassador?'

'How on earth would I know?' retorted the Ambassador; 'I never even knew my great uncle!'

Ambassador von Richthofen, although not averse to media attention soon tired of the questions and turned to his aide for an exit strategy. He thanked Robin for showing him in detail the aeroplane his forebear would forever be associated with;

but now it was time to lose the press and go. After all, how could he answer for the thoughts of the Red Baron – a man he'd never met; a man who, in his brief life, had been an enigma even to close family and fellow aviators who'd fought alongside him.

With the press gone it was time for the flying display, and so Robin took his place at the controls and fired up the engine. This was what Ambassador von Richthofen had come to see and Robin at the controls of the triplane was not about to disappoint.

Separation

It has been said and also written, that men's dedication to their flying machines can leave them separated or divorced from their partners, and that as a breed they can be at best: "detached" and at worst "deranged".

Or: "a weird lot whose obsession might be deemed close to clinical madness."

That flying is a drug that requires its daily fix and like all drugs leaves the addict hopelessly dependent.

The point had been reached – after plenty of warnings, threats to leave, etc, that separation was the only way forward. Better to go now than let it drag on week after week, month after month, year after year. Wendy could stand it no longer, and so she packed her bags and returned to her parents' home in Plymouth.

Enter Howard and Sheila Truscott

Howard and Sheila Truscott, had bought an old mine cottage– the old pay office for the tungsten mine at Hemerdon Ball on the edge of Dartmoor and it was from here that they first saw Robin Bowes taking part in hill climbs in his E-type

Jag. The Truscotts became even more acquainted with Robin when Howard and Sheila – along with their two small daughters Sarah and Helen – moved down to Newnham Road in Plympton.

By 1975 Robin was self-employed and working on his own selling cars from his new home in Newnham Road. Sheila would often see him on the driveway preparing a car for sale as she walked the children into town on a shopping trip. Howard had set up his own car body repair business and Robin would occasionally bring him custom especially with the stock cars he was racing; so over time they were getting to know one another quite well. Howard and Sheila were themselves keen race enthusiasts and would watch Robin competing at Oddicombe and Babbacombe.

It surprised Sheila that Robin had another facet to his character: he was a drummer in the Rod Mason band but to Sheila he didn't appear to be the typical drummer type as personified by people like Keith Moon, Mick Fleetwood or "Animal" from *The Muppet show*. Robin rarely performed a drum solo and he didn't seem to fit the manic profile; it seemed to her that he didn't do anything unless he was going to do it right and that whatever he put his hands to he was good at.

When the separation came, Wendy left Robin with the Old Inn House, but it was necessary to split everything down the middle in an amicable arrangement. She needed her half of the proceeds including the value of the aeroplanes and so Robin approached Howard and Sheila and asked them if they might like to buy half the property in addition to renting the large garage for Howard's business. After all, the house was big enough – Robin could have the ground floor leaving the upper floor to be rented out. Would they like to live there?

A family of four in just the top of the building might be a bit of a push, and with the girls needing to be ferried by car – usually into Plymouth city centre on a regular basis, then

Ermington was just that bit too far out of the way for practical purposes. Better to stay in Plympton, but yes, the idea of buying into the Old Inn House and renting the top half was a good plan. Furthermore, the idea of Robin and Howard becoming business partners in the garage seemed a natural progression in their relationship. Robin would buy and sell cars while Howard concentrated on the body repairs. All round, it seemed a really good solution for all concerned.

Howard and Sheila would rent out their share of the Old Inn House, which would enable Robin to remain in the home he'd come to love. Howard's former business partner would continue to run the business that they'd set up in Plympton, which meant that their clientele – built up over many years – would remain with the old business.

The Truscotts also set to work making the upstairs of the Old Inn House self-contained for bedsitters – making four bedrooms in total. Over the mews entrance there was a large room that was converted into a kitchen-cum-living room for the tenants, and there was a shower and a bathroom, and a smaller shower room. Robin was happily accommodated downstairs whilst Sheila commuted from home in her new role of cleaner, tidying up for what was now a house of three occupants and made up of two tenants: Andy plus Jason the apprentice both employed by Howard for the business; and display pilot and sometime car salesman, Robin Bowes – mightily relieved to be saved from eviction.

A new little community had sprung up at the Old Inn House and Ermington had its very own car body repair business.

Chapter 8 – Ireland: The Ulster Air Show and Air Spectacular

Robin was always very aware that when flying the triplane he was piloting a machine that had been designed a mere 14 years after the Wright brothers first flight. People would ask him, whether he had ever been scared whilst flying the triplane. He would reply in understated fashion: 'There have been moments of concern', but those moments of concern were for the aeroplane, not for his safety. He felt that he would never be able to forgive himself if it was to be accidentally damaged.

One thing that did concern him, however, was that one day, through some reason whether medical, financial or any other, he'd have to stop flying it and everything would become a memory.

Now that really did scare him.

Farewell to Airymouse

One of the things that had to go to cover the costs of the separation from Wendy was Airymouse, the little biplane made famous by Harald Penrose. Robin kept it at Dunkeswell and it had even featured in a TSW news report with Robin at the controls and also Dave Silsbury, Robin's good friend who brought so much of his skill with woodworking to repairing and rebuilding aeroplanes like Airymouse.

Robin, a talented engineer in his own right, was also a frustrated writer. He had hoped to write of Airymouse as Penrose had done in his famous book *Airymouse*, but the opportunity to write anything more than a magazine article never quite materialized. The article, published in the Popular Flying Association magazine, had at least helped to put him in touch with Penrose.

Those aviation fans who knew the famous story recognised Airymouse at every event Robin was able to attend with the Curriewot.

'Is this the same Airymouse that Harald Penrose wrote about?' they would ask.

'Yes, the very same,' replied Robin.

Robin knew from experience that their questions concerned the aeroplane and not the pilot and so he never let it go to his head for one minute that any of the admirers were interested in just what the pilot's life was like.

John Currie had called his little everyman biplane the "Wot" because so many people had asked, 'What is it? What are you going to call it?' Now, people knew exactly what this Wot was – it was "Airymouse" and it was very famous indeed.

A few years earlier in 1985, the year that Robin bought Airymouse, he'd received a call from Jeff Salter at Newtonards in Northern Ireland. Jeff was an Air Traffic Controller at Aldergrove Airport, Belfast, and the airport was celebrating its quarter century with a first ever flying display.

Jeff acting as liaison between the Air Traffic Control Organisation and the Airport Authority called Robin at home in Ermington in reply to concerns regarding the triplane's requirements on arrival at Aldergrove. Prior to the flying season such enquiries had to be made: "Was the grass strip adjacent to the main runway? What accommodation is available?" All the usual enquiries that Robin had to make each and every season. Jeff had not expected to be on the phone long, but in total the call came to an hour. The two men hit it off straight away and spoke at length about all manner of things involved in getting this historic aeroplane all the way to Northern Ireland. Jeff was fascinated by the triplane and like so many of his colleagues around the UK was looking forward to actually seeing it – a flying machine of legend no less.

To get the triplane to Northern Ireland was a major feat of logistics requiring Robin to set off from Dunkeswell on the Thursday prior to a Saturday show. Because the triplane had no brakes or tail wheel, he could only put down at aerodromes that could provide a safe grass landing so planning for his route north had to be meticulous. Crossing the Irish Sea was the most hazardous part of the journey for if the engine was to quit over water the result could be catastrophic, so the shortest crossing was always preferred.

On his very first trip across, RAF Valley, which was base for the first leg of the journey, equipped him with a PLB (Personal Locator Beacon) and a Life Preserver with the proviso he return the items on his way home and that he attend their display later that year. But from Anglesey the trip to Northern Ireland is the longer trip across water and so subsequent flights were more easily done with a stop off at Manchester's Barton Aerodrome, which was very General Aviation (G.A.) flight friendly. Only on later trips did he go via Carlisle and the Solway Firth to Stranraer for the shortest water crossing at just over 22 miles.

Robin also had no modern navigational aids for such a large stretch of water – simply a map and a compass. Trying to hold a map in an open cockpit is a feat of endurance in itself coupled with trying to keep the machine on a steady course as the tail plane needed constant pressure from his feet; there being no fixed fin area to allow him the chance to relax his feet away from the rudder pedals.

The trip home also took two days, so to display in Northern Ireland he had to allocate almost an entire week. It made sound economic sense when bringing an aeroplane so far, to display at more than one show; and as the vast majority of display aircraft, especially the warbirds of WW2 and the First War replicas were nearly all based on the English side of the

Irish Sea, it made sense for the organisers to liaise on dates. With the success of The Ulster Air Show came other shows at Enniskillen and Bangor held on subsequent days and with no charges for aeroplanes that had to be left on the tarmac for any length of time.

The only thing that could stop a trip to Northern Ireland was nature itself in the form of a strong headwind and it stopped Robin just the once on his way to display the triplane at a smaller show Jeff had organised near Londonderry and sponsored by a shoe manufacturer. Robin was coming especially for this little show but couldn't get beyond Blackpool. He called Jeff and was very apologetic: 'Jeff, I'm not going to make it tonight, I'm sorry. I rarely ever don't make it, but I've got something like a 25 to 30 knot headwind. If I go round the Irish Sea, I'll be too late; if I go straight across I'll run out of fuel before I get to you.'

The weather could also scupper an air show and for a time it looked as if it was going to scupper that very first show at Aldergrove. The rain had poured all through the previous week and the build up of surface water on tarmac and concrete was causing serious concern among the organisers. Inclement weather is the bane of show organisers and pilots alike with low cloud and summer showers resulting in the cancellation of many an event.

Miraculously, on the morning of the show a meteorological respite was in the offing. As Jeff waited anxiously at Aldergrove, an early morning mist was causing trepidation for all those taking part. The triplane had been accommodated overnight at a small farm strip in County Armagh where there had been an available hangar. Now, flying with two escorting private planes from the strip, Robin was guided toward Aldergrove. Out of the early morning haze Jeff could make out a formation of three aircraft approaching – two modern single-engine aeroplanes and the remarkable Fokker triplane – the apparition of which was not unlike that of the ghostly *Flying*

Dutchman emerging from the mist.

Triplane in flight (credit Paul Harrison Photography)

The grass was so wet that as Robin touched down a vast flume of water burst all around the triplane as if it were a fairground water splash ride. But at least Robin had arrived safely, his epic journey in so tiny and vulnerable an aeroplane was over; and fortuitously, the sun was now breaking through. Surfaces were quickly beginning to dry just in time for the crowds that would soon be arriving.

If Robin was by then an old hand on the air show circuit, Jeff Salter was experiencing it all for the first time and as an organiser no less! Even as a child growing up on the South coast of England he'd never been to an air show before and so had no yardstick by which he could measure developments. For his first Ulster Air Show at Newtonards the weather was on his side, but it took him 45 minutes simply to walk from the control tower to the corner of the hangar because every time he took a pace somebody would approach and ask a question. While he was talking to them, three other people stood by patiently awaiting their turn.

Robin's advice was "delegate"; and, over the coming nine years, the two friends would help and advise each other whenever there was a need. Jeff would sometimes act as a wing walker for the triplane, but he soon discovered that this could be a bruising task not to say exhausting as the "tripe" taxied along at a cracking pace. Holding the inter-plane strut at arm's length and at shoulder height was particularly precarious especially if a wheel dropped into a rut the middle wing would instantly drop before riding up and whacking him mighty blows to both the shoulder and head on both the down and upward movements. To turn the machine fully required the wing walker to dig their heels in and use both hands to hold firm and retard the wing in order for it to turn into wind.

Robin, for his part, would help with any on the spot organisation that might have been needed such as sorting out a new running order for the display when bad weather came in and threatened to ruin a show. Together with fellow display pilot, John Romain, a running order was worked out that took the last half of the display and ran it in front of the slot allocated to the Red Arrows meaning that the Reds would not be delayed and could come in on time oblivious to the fact that what would have been the following displays had all been brought forward so beating the arrival of the incoming storm.

The troubles in Northern Ireland did not affect the show until such time as the Ministry of Defence were given intelligence that the IRA possessed Surface-to-Air missile launchers. Jeff and the organising committee had been determined that shows would go on despite the troubles and that would mean bringing in the RAF regardless of MOD reservations. The Ministry for its part, had decided – without informing show organisers – that non-operational flights over Northern Ireland could not be permitted due to the threat of a S-A-M strike. This was also the first year that the BBC was to broadcast the show for its Ulster viewers and ironically it was through the media that the public became aware of the ban. A mighty fuss ensued with politicians from all parties becoming involved and such was the strength of feeling that Whitehall was forced to reconsider and consented to a solitary Hawk being sent across.

In just a few, short years the air show at Newtonards had become an extremely popular event with both the public and the pilots. In the midst of a troubled period for the province the public couldn't get enough; it was a very welcome event. Pilots were treated like Hollywood stars. Two Swedish Air Force pilots who'd arrived in Ulster flying a couple of Drakkens were practically mobbed when they walked along the crowd-line. Such a sight had never been seen in Ulster with spectators begging them to stop and sign programmes, which they happily did, even though no one in the crowd could actually say who the pilots were. They just knew that pilots were "stars".

The pilots, too, loved the welcome they received from both the public and the show's organisers and would move heaven and earth in order to attend. They were looked after in every respect from the supplies of fuel required to food and accommodation; nothing was left out, no question went unanswered, and no one was ignored.

Newtonard's reputation was spreading far and wide and before long pilots were rating it in the top five of British air shows.

The BBC coverage of the air show not only helped to boost attendance, but also highlighted the work of both military and civilian display pilots like Robin Bowes. Seasoned aviation commentators added their voices to the event and regional presenters hammed it up for their magazine programme audiences.

Technical innovations in broadcast cameras were also making it possible to shoot footage from the aeroplanes with close up views of the pilots and what they could see ahead of them. Pilots Mark Hanna and Nick Gray recreated a Battle of Britain chase sequence with mini cameras fixed inboard the cockpits of their respective machines – a Spitfire and a Messerschmitt 109. Pre-recorded prior to the show, the footage captured a pilot's eye view as he closes on an adversary and for its day, this technique was extremely innovatory.

It IS a Fokker Triplane!

Robin's flights across country never failed to attract attention from those air traffic controllers who could hardly believe their ears when he identified himself as a "replica Fokker triplane". Flying over the English Midlands, one military base controller was only convinced by the eye witness report of a Harrier pilot who, using the Harrier's variable thrusters slowed to the triplane's tortoise like cruising speed of 85mph before confirming Robin's identification.

'My goodness! He's right! It's a Fokker triplane!' radioed the astonished pilot as he pulled alongside its engine roaring and drowning out the piston chatter from the old Warner Scarab.

On later visits across the Irish Sea, Robin, not content with displaying just the one aeroplane, actually ferried three for display at Newtonards – the triplane, a Pitts Special (that he'd bought to promote Television South-West and was being sponsored by Volkswagen for the show), and a scaled down replica Focke Wulf 190 that belonged to a friend, Peter Logsdon.

Jeff and Robin had become firm friends, with each being mutually supportive of the other. Robin would go to great lengths to help make the now annual show a success. Jeff, as one of the organisers, had to be careful how he rostered the three performances on the day of the show as no sooner had Robin flown one aeroplane and climbed out of the cockpit than he had to prepare the next for display.

When not flying a Crunchie Stearman or his own machines, Robin would fly aeroplanes owned by other enthusiasts such as this 2/3rds scale replica Focke Wulf 190 belonging to Peter Logsdon.

By far, the biggest problem was ferrying all three aeroplanes single-handed. This laborious process involved catching commercial flights from Plymouth to Heathrow and from there on to Aldergrove – a total of twelve separate passenger trips. The entire venture took over a week to complete, but Robin thought it more than worth it to please the Northern Irish crowds.

On his final trip through Heathrow en-route for Aldegrove, he was even carrying a replacement propeller made especially for Airymouse by Dave Silsbury. Dodging around the suited business commuters, he approached the British Airways check-in desk carrying not only the prop wrapped in brown paper and cardboard, but wearing his flying suit and Red Baron fur lined leather jacket plus a life preserver and flying helmet as he had no luggage capacity aboard the triplane he had to wear what he needed for the homeward journey.

The young woman at the check-in desk put down her pen,

looked him up and down and then asked with a wry smile: 'What's the matter, sir? Don't you trust us? You know, we can look after you just to get you to Northern Ireland.'

Robin thought this was highly amusing.

Ringing the Changes

Jeff had been looking for some time for a light aeroplane that he could own. An experienced microlight pilot, he'd asked Robin's advice about what sort of aeroplane he could buy with a limited budget quite unaware that events at the Old Inn House would result in Airymouse having to go.

Robin knew that something had to go if he was to retain the house. The only proviso that he made in selling Airymouse, was that should Jeff tire of it and wish to sell it on could he please have first refusal. Jeff knew just how fond Robin was of the Currie Wot and was surprised that he was even thinking of selling it especially considering its history and associations with the Westcountry. He suspected that financial problems were the root cause, but Robin, always discrete, would not elaborate on just why Airymouse had to go.

Over the course of a decade, Robin brought his aeroplanes to Northern Ireland as a regular contributor, but he knew that attending every year could be counter productive. The crowds would expect change and the changes had to be rung every now and then.

He understood well this principle that year after year after year he couldn't keep on bringing the same aeroplanes. That was one of Robin's great strengths and just one of the reasons he was so welcomed on the other side of the Irish Sea.

Triplane bookings. Map showing all the air show venue bookings for the triplane during the 1988 season in England and Ulster. (The UK hosted more air show events than any other country in Europe.)

Dublin

'I was flying along at a thousand feet and there below me I saw the bright red triplane – it was the Red Baron. I dived down behind him, got him in my sights. He tried to dodge along a riverbed and through some trees, but he couldn't escape. I opened up my twin machine guns and the black smoke poured from the back of his aircraft. I had shot down the Red Baron. And then, through the earphones, came the voice of the film director: "Okay Ken, that looked good, but just one more time." Four times that day I shot down the Red Baron.'

Ken Byrne, Irish pilot, who flew in WW1 films shot in Ireland – films like *The Blue Max* (1966), *Darling Lily* (1969) and *Von Richthofen in Brown* (1971) (US title: *The Red Baron*). The Red Baron featured in all three films.

'Why not run an Air Show?

We could have fun at the same time doing something we enjoy,' suggested Madeleine O'Rouke to the Irish Aviation Club at a monthly meeting in 1977. She was secretary to the Dublin Ballooning Club and the discussion that evening was how to raise vital funds for the IAC, which had charitable status. A raffle they had organised had actually cost money and there was a bad case of raffle fatigue amongst the stalwart supporters. Coffee mornings and cheese and wine parties made good promotional events but were of little benefit for raising money.

With the best intentions in the world, suggestions have a way of coming back to bite the "suggestee" and for the next ten years Madeleine's calendar dates were circled full of organising appointments as she – as event secretary – and a highly qualified committee of fellow aviators including an airline pilot and Air Traffic Controller gave up holidays and free time to organise the first Air Spectacular.

Fairyhouse racecourse in County Meath was chosen as the

venue as it was rural, flat, had ready-made facilities, and could easily accommodate spectators and their cars. Local aero clubs had used the course for training and it had been licensed for use as an airstrip since 1965. All that was needed was a permit for the show.

Organising was a learning curve for all involved as new skills had to be mastered such as marketing, traffic planning, catering, and liaison with official bodies such as the Garda and local councils.

Madeleine's remit included finding acts for the first show and that meant looking across to the UK where the bulk of display pilots resided. The UK hosted more air show events than any other country in Europe.

This was a great opportunity because Madeleine could invite aviators not normally seen in Irish skies – people like Brian Lecomber, the author and civilian display pilot. The only drawback was the limited purse – in this case a mere £500; and Brian's fee would need to be at least twice that to come the distance with his Stampe.

Lecomber was to become the first aerobatic pilot to be booked, and Fairyhouse was to become one of his favourite shows. He thought it a superb venue for what was then only a small display out in the sticks. Flying an orange and blue Stampe, his was the opening act of that very first show on the 27th of August 1978.

In its first three years, Air Spectacular established a remarkable reputation with spectator attendance rising from 9,000 to 30,000 by 1981. The partnership between the now Irish Aviation Council (formerly the Irish Aviation Club) and British Airways came to an amicable end. BA just didn't have the number of staff in their Dublin office to cope with a show that was growing so rapidly and therefore Aer Rianta filled the requirement for a main sponsor.

Although Air Spectacular '84 was a successful show, it was

also marred by tragedy when a Polish pilot, Jan Baran, crashed in his Wilgas high wing monoplane just five minutes into an aerobatic display with his fellow team members. He was part of a small, but award winning Polish team of display pilots who were touring Europe, winning trophies and admiration everywhere they went.

Baran was killed on impact and there was nothing anyone could do to avert the disaster. However, the show organisers – many, experienced aviators themselves – had considered the possibility and carefully planned for such an incident by bringing in and having ready the very best fire and rescue team so that it was assembled on site and prepared to go at a moment's notice. Such preparation meant that crash tenders were on the scene within a minute of impact breaking through plastic fencing as they went and dowsing the burning plane in foam.

The show continued that day not through any lack of respect for the loss of a great pilot, but through the necessity, as Madeleine later recalled in her book *Air Spectaculars* of keeping "exit routes clear for ambulances and emergency services." If the crowds were to leave en-masse then the roads beyond the perimeters would soon be blocked. For air show organisers, this is a common and universal contingency practice should a tragedy such as this occur.

The following year, for Air Spectacular '85, the opportunity to move away from Fairyhouse racecourse was taken, not because of the previous year's tragic accident – as that was no reflection on the venue or the organisers – but the invitation was received from sponsor Aer Rianta to tie in the Air Spectacular with Cork's 800th anniversary celebrations; it would be the airport's contribution to the festivities.

This seemed a marvellous opportunity to take the show to the South coast and thus enabling a whole new audience to see an even greater display of aviation talent. For the first time in its short history, Air Spectacular could be held at a purpose-built airport, which in itself would enable a far greater participation of entrants from around the world – both civil and military. In addition to the Irish Air Corps, French, German and Italian air forces were sending contingents and the Americans would display F-15, F-111 and F-16 fighters as well as the HH53 Jolly Green Giant transport helicopter, which had been very popular with the crowds in previous years.

The organising committee were experienced enough now to feel the confidence of being able to pull off the biggest and best show yet. It was a fantastic opportunity. Even national broadcaster RTE was on board with the aim of producing a half-hour documentary on the Cork Air Spectacular hosted by Pat Kenny.

But with all the best preparations in the world, the most meticulous organisation cannot stop the air show organiser's worst nightmare – weather. Within 24 hours of the show getting under way, rain and fog was hampering the arrival of pilots such as Richard Goode flying his own Pitts Special and approaching from Belfast after displaying there. He was well and truly lost, whereabouts unknown and caught in some of the worst weather he'd ever experienced – weather that was to force him to land in a field in the Blackwater Valley of Fermoy some 40 miles north of Cork City. Thankful to be alive having narrowly missed power lines, he literally had to wait it out.

To make matters worse, sea fog was rising to meet the low cloud and with Cork Airport situated at just 500ft above sea level the elements were conspiring to scupper the show. The final weather forecast that was to come too late for any contingency plans that might save the day, was that the cloud base was to remain below 200ft leaving no other option but to cancel at the last moment, which in itself was to be a Herculean task. And it pleased no one at all to realise that ironically, Dublin and Fairyhouse racecourse in particular, were bathed in fine summer sunshine that day.

It was a deep disappointment for all concerned, as a year of considerable hard toil and effort was now wasted and worst of all, the cost of cancellation meant there would be nothing to show for it. It was a huge financial loss for what had been a £30,000 investment in what should have been the greatest Air Spectacular ever seen in Eire.

The immediate post show committee meetings were on this occasion fraught and unhelpful. A cooling-off period before getting round the table might have helped. After all, no one could be blamed for the weather; it was simply a natural part of an Irish summer and was beyond anyone's control. Nonetheless, things were said that shouldn't have been and it was awhile before planning could begin again in earnest for Air Spectacular '86.

Air forces and civilian pilots have to be contacted in advance if they're to make a commitment to attend an air show. The US Air Force actually has a selection committee that consider how best to direct their machines and manpower whose primary role is training for combat and tactical air defence. There is no dedicated display team in Europe unlike the RAF's famous Red Arrows whose raison d'etre is purely public relations and the air shows.

Large support teams normally come with visiting military aircraft bringing spares, equipment and even their own fuel, so it isn't simply a case of requesting an available F-16 or F-111 for a flypast. It's often a lottery as to what comes and is very dependent on what aeroplanes and pilots are available at a particular time with only a six-week notification to the show's organisers of what they can put in the programme.

Civilian pilots come at great personal cost and when there is a cancellation as there had been at Cork, the cost for them is considerable. Veteran pilot Ken Wallis had come all the way from Norfolk with his "Little Nellie" the gyrocopter made famous by the James Bond movies only to be stood down by the weather.

When finding suitable acts from the civilian contingent, organisers like Madeleine O'Rourke were heavily reliant on reputations. She would go to great lengths to ascertain that a performance was going to be both entertaining and safe. Word of mouth and recommendation was crucial as not every act was of a high standard. Some display pilots were not asked back a second time and a poor reputation would follow them. And so the network of air show organisers in Ireland, the UK and on the Continent helped to keep the good pilots employed and keep the cowboys away from the arena. This system didn't always work and sometimes it was as well to check things out first hand.

At her own expense, Madeleine together with her family would travel the length and breadth of the UK visiting air shows and checking out the acts on display. It was a good exercise for talking to pilots and getting a look at how other shows were organised.

One such visit to RAF St Athan in South Wales in 1984 enabled her to carry out a feasibility study, as the show was roughly the same size as their new venue-to-be: Baldonnel. It was here that Madeleine first spotted the red Fokker triplane parked in the viewing enclosure together with its pilot who was tinkering with the machine. He seemed approachable and now was the best opportunity to engage with him and sound him out. She pushed against the barrier and tried to catch his eye, whilst trying to size him up. Would he respond? How would he react?

He turned and came over smiling and they introduced themselves. Madeleine realised with gut feeling that this was a like-minded guy who liked a challenge; he was a true enthusiast with a glimmer in his eye. It was going to be a hell of a long way to come, but he was positive.

'Can you bring the triplane to Air Spectacular at Baldonnel as the theme is going to be WW1 aviation?'

'I'd enjoy that!' he replied, and so a provisional booking was made for Air Spectacular '86 Baldonnel.

Chapter 9 – Barnstorming & Test Flying

When regional broadcaster Television South-West approached Robin for advice on having a hot-air balloon that they could use to promote their logo at events, he suggested they consider an aeroplane instead.

'Think about it,' he said. 'When do you normally see a hot-air balloon? Early morning or late evening when the air is still and it's a clear sky. How many events are going to enjoy those conditions? And the balloon can only follow the warm air currents. Direction is always a problem as is collecting the balloon when it's down.

'Why don't you have an aeroplane? You can display your logo on its wings and fuselage, and you can probably incorporate the logo into the registration and fly at anytime of the day in all but low cloud.'

Due consideration was paid to Robin's suggestion and a tentative toe-in-the-water contract was drawn up. No one at TSW was going to make any promises about the years ahead, but at least for the 1989 season they would give the aeroplane idea a go at least. And it gave Robin a promise of some income for the nine months of the first contract running from May 25th to December 31st.

Robin suggested they have an acrobatic aeroplane such as a Pitts Special. However, the capital investment for a Pitts was going to be considerable; it wasn't as if TSW were offering to buy the plane and employ him to fly at their cost. The burden on purchasing, maintaining and insuring would be his alone. The outlay for such a high performance aeroplane was a crippling £28,000.

With more time, it might have been possible to purchase the

plans from the US and construct a home-build Pitts, as many of the type had been built in workshops by privateers. This would have been the cheaper option, but it would be impossible to have the aeroplane ready for the season.

At least the purchase of a ready-made aeroplane was a value that was not expected to depreciate in the coming years and it was, as Robin pointed out to his bank manager, "the best tool for the job."

At Land's End aerodrome, word had it that Bill Penaluna had built a Pitts Special SI-S registered G-BLHE. This was an ideal opportunity.

Contracted to fly 20 displays of their choosing between July and the end of September 1989, TSW would pay £10,000 for a season's display flying – £6,000 at the commencement of the contract with further payments to be made at the time of the first display; the final payment after ten displays plus costs for the aeroplane to be sprayed in their brand livery of green and white with logo and registered with the CAA as G-OTSW. In addition, they would pay £375 for each display. Robin, as owner/operator was free to use the Pitts at any other event of his choosing.

Robin estimated the operating costs for a season including insurance to be in the region of £3,500. The TSW region he was expected to cover was a broad area stretching from Weymouth in the East to Land's End in the West, so travelling to shows was likely to be a costly business both in fuel and time.

To raise the remaining finance to purchase the Pitts, Robin suggested to his bank manager that he could if needs be sell other assets in his collection such as the Hawker Nimrod replica and an EAA biplane that he was restoring at Ermington. Each was capable of fetching between £12,000 or £13,000 should they be put on the market.

The Pitts Special SI-S was far more than just another biplane for Robin. To many civilian pilots, the type represents the ultimate challenge in flying skill and endurance. A pilot doesn't have to be young to fly it, but it's vital that they be physically fit as the G-forces on the body can range from +9 to –6G. Such heavy gravitational forces are particularly severe if the aeroplane is to be flown in the dynamic ways its designer (Curtis Pitts) intended.

With TSW on board as the main sponsor, Robin's bank manager agreed the loan and so a second Land's End purchase was added to B.C. Flight.

Robin planned a routine that would show off the Pitts whilst giving ample exposure to the TSW logo sign-written on the top wing surface and under the lower wing in big black letters. The sequence depended on the conditions for the day of the display, wind strength and direction, cloud base and the position of the sun not to mention what Robin had had for lunch.

A dynamic entrance was crucial for catching the crowd's attention. Dive in at 190mph, then climb vertically showing off the wing logos. Quarter roll and a stall turn at the top gave the photographers with the best lenses a chance to focus in on the flying billboard. From the stall turn, downward three quarter roll to Cuban eight, then five 'G' to negative. Stall turn to hesitation roll, climbing three quarter roll with turn around to barrel roll; knife edge pass climbing 180 degrees turn to loop, climbing and rolling 180 degrees, turn for the dive followed by a three quarter upward roll then tumble at the top. With nil airspeed the Pitts revolves around itself. Dive out for turn around to inverted pass for the crowd with push around 2- 4 G. Roll on descending 45 to stall turn, loop to stall turn, hesitation roll to turn around for inverted rock pass. Then a final pass with canopy open waving frantically with the encouragement that the enthralled masses might just wave back!

The display worked particularly well where the Pitts was the sole flying attraction such as fetes, carnivals, country shows and regattas.

1989 was a good opening season for the Pitts and 1990 was promising as there was another use for such an aircraft that neither Robin nor TSW had initially considered: that of traffic spotting. It was at the 1990 Devon County Show at Exeter that he was able to pass on valuable traffic reports to the organisers as the M5 slowed to a standstill. And at the Royal Cornwall County Show at Wadebridge the Pitts proved to be the ideal vehicle for air-to-air filming for the TSW *Today* magazine programme.

The Broadcasting Act of 1990 was to throw a serious spanner in the works of regional ITV broadcasters. Sponsorship with TSW might have continued for a third season had it not been for the interference of Prime Minister Thatcher's Conservative government as they sought to shake up the broadcasters by bringing in franchise renewals for the ITV companies. Contracts for new regional programming were put on hold, as no one knew just what amount of money would be needed to renew the franchise.

It was a fiercely controversial move by the government on regional broadcasters that seemed to reflect personal antagonisms rather than a straightforward round of franchise renewal. TSW appealed with a judicial review against the ITC and it even went as far as the House of Lords only to be rejected.

Robin also appealed only in this case to TSW in the hope that they might find a way through the situation and agree to some degree of sponsorship. Over two seasons, the TSW Pitts Special had successfully built up quite a following, and it was the branding that people recognised; to lose it now, argued Robin, would be to lose everything that had been built up. It also meant a significant loss of income whilst the continuation of a loan on a Pitts Special was still to be paid off. He hoped that other ways could be found of keeping a limited sponsorship going or even using it as an airborne camera mount for events and reportage; but with the franchise renewal looming all departments at TSW were faced with considerable budget cuts as the Chief Executives sent out orders to rein in spending.

All to no avail

The loss of contract was to pull the rug from under Robin as he was preparing for the third season and just as he was beginning to get used to the regular work and income, not to mention exposure. He was not simply the man who flew as the Red Baron, but the highly able acrobatics pilot who flew the TSW Pitts Special. To appease his bank manager, the Pitts would now have to be put on the market.

Awards & Trophies

26.11.90

RAF Benevolent Fund

Dear Mr.Bowes,

Richard Roberts the honorary secretary of the Association of Air Display Organisers and Participants has sent me a cheque for £250, which he says represents an award made to you for the most meritorious air show during the past season. He tells me that you wish this to be passed on the Royal Air Force Benevolent fund as a charity of your choice. We are most grateful for this very generous gesture and I want to pass on to you the very sincere thanks of the chairman and council of the benevolent fund ...

Edgar Hamilton Ltd.,
Great Eastern St.,
London,

Dear Mr.Bowes,

I thank you for your letter of 12th of November 1990. Unfortunately I was not able to see your display this year, but look forward to rectifying this in the 1991 season.

Your generosity in donating our reward to the RAF's Benevolent Fund's Battle of Britain 50th Anniversary Appeal is very much appreciated by all of us here and I cannot think of a more deserving cause.

I look forward to meeting you in the near future.

Yours sincerely,

…

Trouble at Mill

There were times during their business partnership, when Howard Truscott would get quite irate with Robin because he was hardly ever in the workshop, especially in the middle of the summer when Robin was flying midweek at shows such as Culdrose and St. Mawgan; in addition to the weekend shows for which he'd frequently disappear on a Friday afternoon and not return until late Monday. Tuesdays might be a press day for the midweek events and then there were the long distance events in places as far afield as the Irish Republic and Belgium for which he would be away for the best part of the week. Quite understandably, Howard was beginning to lose his patience.

'I nearly killed old Bosey today!' exclaimed a very irate Howard as he sat down for his evening dinner with Sheila.

'My God! What happened?' asked Sheila.

'I had my hands around his throat and I'd have throttled

him had Mark Parmenter not turned up and pulled me off him.'

For Howard, Robin was spending too much time ferrying aeroplanes to shows and attending press days while hardly anytime was being spent on the business of selling cars. Howard's frustration had been brewing for some time and now it had reached boiling point. Had it not been for the timely intervention of one of Howard's biker buddies – off duty motorcycle policeman Mark Parmenter, who had just taken a detour to drop in on Howard – the episode could have taken on a more serious note.

Howard – a hard worker at the best of times – was at his wit's end. The partnership seemed nothing more than a one-sided arrangement; and much as he liked Robin and admired what he did, the practicalities of running a business demanded a much greater commitment if it was to work.

That afternoon, things had come to a head under the archway of the Old Inn house and by the time Mark arrived on the scene Howard had pinned Robin to the wall with both hands around his neck. It wasn't as if Robin himself wasn't working hard because he was – *extremely hard*; it just wasn't centred on the business he shared with Howard.

The partnership had begun some four years earlier at a time when there hadn't been a demand on Robin's flying. A few displays per season – half a dozen at most – was how it had begun with the triplane, but now the commitment to attend displays with the Pitts, triplane, Nimrod and Airymouse was taking up the bulk of Robin's time.

Without meaning to upset his friend and business partner, Robin had neglected his original agreement with Howard as the demand for flying at displays had increased. And so it was agreed to dissolve the partnership with no hard feelings. Relieved of their mutual commitments, the friends were able to resume an otherwise affable relationship.

It wasn't just the breakdown in the partnership with Robin; Howard didn't want to do car repairs any more. He had been paint spraying for much of his working life and although over the years protective equipment had vastly improved – and he had breathing apparatus if he needed it – his lungs had suffered the effects of inhaling too much cellulose which was causing him frequent coughing bouts; yet he'd never smoked and rarely drank alcohol. He was sure that only a change of business would save his health and sanity, and as he approached 43 it seemed a good opportunity to stop. Following the break Robin had bought back The Old Inn house although Howard continued to rent the large shed. He would do all his own work and if Robin wanted any work done Howard would charge him at a discounted rate and vice versa. And that system seemed to work much better because when Robin was away flying it didn't matter to Howard. He didn't feel that Robin owed him time.

In July 1991, Robin displayed the triplane at RAF Chivenor and the following month at RAF St. Mawgan. Howard, who had set up a new business with a simulator ride at Paignton Zoo, took the rare opportunity of attending the British motorcycle grand prix at Donnington Park and loved it. Returning home late in the evening he retired to bed happy but exhausted; it had been a long day's riding from Devon to the Midlands race circuit and back again.

Suddenly, settling into bed, he told Sheila that he was getting numb sensations in his foot and hand: 'Oh, blimey! Don't switch the light off yet. I've got pins and needles.'

He tried to get out of bed but collapsed falling between the bed and the wall.

'What the hell are you playing at?' asked Sheila as she desperately tried to pull him back onto the bed.

'It feels like pins and needles, but I don't think it is,' he told her.

She called the doctor immediately and as his speech began to slur he urged her to call their friends John and Margaret so that they could look after her.

Howard died later the following day in hospital having suffered a severe brain haemorrhage. He was 42 and had been married to Sheila for 21 years.

Robin had been flying the triplane at the Sanicole Air Show, Hechtel, Belgium and returned to the shocking news on the day Howard's life support system was turned off. Howard had enjoyed largely good health and there had been no indication that any problem existed that would cause his sudden demise.

In a rare moment of intimacy, Robin gathered Sheila up in his arms. His gesture surprised her as he was not normally demonstrative or tactile and the comforting embrace was so welcome. They were both now without their partners.

In the weeks and months that followed, they recalled happy memories of experiences shared at the Old Inn house. They recalled how on working days, Robin would come down from the garage when Howard had his lunch. Sheila would come up from their home in Plympton having baked cakes during the morning and they'd sit and talk about all manner of subjects over tea and cake in the kitchen of the Old Inn house. If Howard was working particularly hard there wasn't time for a lot of chatting, but lunchtimes were always sacrosanct.

Robin desperately wanted to do something positive for Sheila in memory of Howard, and so following the funeral they talked about where the ashes might be scattered.

In life, Howard had hated the idea of being cooped up – he certainly didn't want to be put in a box in the ground. Sheila asked Robin if he would scatter the ashes at Donnington Park – if the authorities would give permission, as he'd been so happy watching the grand prix that weekend. Robin suggested they scatter the ashes from the air for which he would seek the necessary permissions.

As Donnington was so close to East Midlands Airport, it was necessary to get special permission from the airport authorities. Robin borrowed a plane, and with three of Howard's closest friends from the BMW Motorcycle Club onboard they took off for Donnington without Sheila, Sarah or Helen who had celebrated her 17th birthday only a week before her father died. Robin requested a pass over Donnington Park – once to check out the lay of the land and the second run to drop the ashes. ATC at East Midlands generously cleared the airspace – a particularly kind gesture for such a large and busy airport.

Disaster Over Germany

June 1992 en-route to Hamm Balloon Festival, the triplane must have caused a lot of anxious glances as French, Belgians and Germans looked skyward to see the Red Baron flying again. They might have wondered what strange ghost was this scarlet apparition in a continental blue sky after all the many years of conflict and division? Was it an omen? Some might have joked that it was a Belgian secret weapon, or that it was all that the Luftwaffe could offer up by way of a deterrent in the late 20th century.

It was, of course, yet another epic journey for the daring pilot who flew as the Red Baron. In his open cockpit, he couldn't relax in the way that a Piper Cub pilot might relax a little on such a journey; he couldn't take his feet off the confounded rudder pedals for one second and he had to keep a constant look out for possible emergency landing strips. This had become second nature flying the triplane; he was always looking to the next horizon, because that's where it might happen – the dreaded failure of the engine.

The Warner Scarab engine had already clocked up some 500 hours in the time Robin had been flying the triplane including an earlier trip to Germany. A historic component in its own right it had powered a Fairchild Argos during WW2.

The Scarab had let him down a couple of times in the past – at Old Warden in the middle of a display and again at Booker near White Waltham when the triplane was being filmed for the BBC documentary series *Reach For the Skies*.

The failure at Old Warden had been as close to a disaster as he'd ever come flying the triplane when the engine cut out at the height of a turn. With considerable ability and quick thinking Robin pulled hard on the stick and glided in to land raising in the process a rapturous applause from the crowd. Stripping the engine later they found that internally the engine was in such a mess it had to be completely rebuilt.

Nor was the Scarab a suitable engine for the design of the replica triplane; its length required plenty of airflow to keep it cool and therefore the cowling had to be longer than its original to allow for greater airflow. This had not been a problem for von Richthofen's Oberursel powered rotary.

Somewhere to Go

In the event of a loss of power a pilot's need to find somewhere to go – to land safely – was an essential preoccupation of flying and always in the forefront of Robin's mind when flying any aeroplane, especially over such a distance.

The perceived wisdom was, look for which way the wind is blowing: are there any bonfires? Chimney smoke? Which way are the cows facing?

Now, at 1200 feet – remaining below controlled airspace – and within thirty miles of his destination, Robin was faced with another Warner Scarab disaster as the engine began to complain with a fearsome offbeat burring noise accompanied by an almighty rattle that shook the airframe. Robin throttled back immediately to reduce the shaking. Now was the time to put that foresight into practice.

Below him, the territory couldn't have been worse for an emergency landing. He was directly over the industrial Ruhr area of North Westphalia – a frequent target for the RAF and USAAF in 1943. Below him was a landscape of factories, roads, rail lines, power lines and schools in the Gelsenkirchen area including the large Aral fuel depot. There was just one possibility presenting itself: a scrubby patch of green field surrounded by trees and hedges; an isolated, overlooked remnant of the once tiny agrarian society that had existed here since the 12th century. It wasn't much, but it was within gliding distance and therefore the only available escape route.

The triplane is built for lift with its triple-decker wing configuration, but it also has three times more drag than a conventional mono wing.

Flying downwind he'd need to turn 180 degrees at 300 feet in order to land into wind meaning that he'd have to lose most of the field in turning. Nose down, struggling to control the ungainly, crippled triplane as it lost height he was rapidly losing ground on which to put down – at least half the length of the field. When he did touch down and connect with the ground in a perfect three-point landing it seemed, for a moment, that he'd successfully rounded out and gotten away with it. However, with no brakes and no steering the tripe was careering toward the edge of the field where, ominously, he could see the top of lampposts, the presence of which suggested there had to be a high bank dropping down to the autobahn on the other side. It was imperative to stop if he was to avoid crashing through the hedge.

At speed, and without immediate action from Robin, the triplane would crash through the hedge and tumble down the bank. He desperately needed to ground loop the careering machine. Pushing hard on the left rudder the triplane slowed, but in ground breaking the side loads were so great as to snap the steel cross wire that braced the undercarriage causing the rig to give way whereby the nose and starboard wings ploughed into the earth breaking and snapping as the aeroplane skidded and shuddered to an abrupt halt.

Tipped up with its tail in the air, but at least not inverted, Robin released his harness and stepped out very carefully for fear of creating further damage to the structure and fabric. He was shaken but not otherwise hurt being well able to walk away with nothing more than a scratch on the leg. Looking back at the wreck from a safe distance, the dream of the Red Baron must have seemed shattered once and for all.

The emergency landing had not gone unnoticed. The field bordered two roads – Ufter Strasse and Grothus Strasse in Hessler. Within minutes a local resident was running across the field to find out what had happened. What a sight! The resplendent Iron Crosses against the scarlet red of the fabric. Could this be happening? The Red Baron had crashed again and he appeared to be all right!

Robin, fearing that he had no German in his vocabulary whatsoever, needed the emergency services informed. He needn't have worried. The sounds of Continental emergency service sirens were heading his way at speed closely followed by the media. Within minutes they began arriving at the field unsure as to whether the report of an aeroplane crash beside the autobahn meant a Jumbo 747 or Bosey in a red Fokker triplane.

It's not unreasonable to wonder that one of those early calls made to the emergency services just might have been along the lines of: 'You're not going to believe this, but the Red Baron has just crashed in a Fokker triplane!'

A local German TV news crew arrived at the scene and began filming the wrecked triplane with particular emphasis on close ups of Robin's reactions. Ever the professional, he fielded questions with the most laudable stoicism. Had he been one of von Richthofen's victims, he may well have been hailed a very worthy adversary before being triumphantly carried aloft to the mess as an honorary guest to enjoy a riotous celebration.

With typical English understatement, he made light of the situation as he viewed the consequences of that wretched engine. The police had cordoned off the crash scene with tape and the chief fire officer was busy liaising with both his superiors and subordinates over a hand held radio. Conscript fire fighters young and gangly, relieved perhaps at not having to fight a fire, now wandered freely around the broken triplane. Their great grandfathers would have known this machine, but to these lads it was an archaic construction of mechanical curiosities – wood, fabric and wires; more like a box kite than a fighting machine. How did this fly from England of all places? And why was an Englishman flying as a German fighter ace of the Kaiser's war?

The English-speaking reporter kept Robin occupied with questions and that was no bad thing as together they contemplated the fate of the triplane. In the warmth of the summer sun, now somewhat overdressed in his green denim flying suit and heavy sheepskin lined leather flying jacket, Robin could easily have been forgiven for breaking down and weeping, his dreams in tatters, or at least appearing shocked and crestfallen as his pride and joy lay collapsed and thoroughly wrecked in a foreign field. And yet here he was managing the situation in the most ebullient manner. Only occasionally, when the jovial banter subsided, did the camera catch an expression of utter desolation as his true concern showed through.

'You're very lucky now?' asked the reporter.

'Yeah, it's a shame such a beautiful aeroplane gets destroyed,' laughed Robin, determined to put his interviewer at ease.

'Isn't it possible to rebuild it?'

'Yes, yes, but it's going to be a lot of work,' laughed Robin.

'Is it your private plane?'

'Yes, yes.'

'Where are you from?'

'I'm from Devon in the South-west of England, near Plymouth.'

'Oh yes, I've been there.'

'I was going over to Hamm. There's a big show there this weekend … and there was only a few miles to go!' laughed Robin, desperately trying to make light of the situation to the young reporter. 'Obviously if you do any press thing, I'd like you to stress that the pilot is okay, because if my friends or relations read it or hear something – they worry.'

'What will happen now?'

'They are coming from the airfield – in Hamm – to collect me, and then I'm not sure. We'll probably dismantle the aeroplane and take it back to the airfield.'

Ironically, in their report of Robin's crash the local German press made no mention of the significance of the triplane nor of von Richthofen the Red Baron. Only that Robin was an "enthusiastic flyer …" and that "[the] triplane [had been] in use during the First World War."

In early July 1917 a similar fate had befallen von Ricthofen, only his experience was through combat. He'd been in a dogfight with a large number of RFC and Fleet Air Arm aircraft including a number of Sopwith triplanes. His Albatross sustained damage that ruptured fuel lines and a partial collapse in the aeroplane's lower wing. Not only was the Albatross badly shot up, a bullet had skimmed von Richthofen's skull with the effect of instantly paralysing and blinding him. As the

Albatross lost height, through sheer will power he battled to regain control of his lifeless limbs and regain his vision.

Below him was a scarred landscape of shell holes and forest; a landing in such a terrain seemed impossible. At least von Richthofen could use the engine to power the Albatross on until his body could stand it no longer and at the limits of his endurance he eventually put down taking some telephone wires down in the process.

Fortunately for the Baron he had landed in German held territory near La Mortaigne thus enabling a safe rescue. Had he fallen prisoner to the Allies at this point the history of the Red Baron might only have run to a paragraph or two. But fortune favours the brave and rather than completing a paragraph the Baron had unwittingly reached the end of his first remarkable chapter.

He had just enough strength to clamber out of the aeroplane before collapsing. His crash landing had been closely observed by his colleagues flying in tight circles over the crash site and by German ground troops who were immediately dispatched to assist the injured pilot and spirit him away to a field hospital at Courtrai.

Surgeons were amazed at his determination – the bullet having skimmed his skull. Following recovery, he could so easily have bowed out of the war a hero; but the determined fighter ace had a date with destiny and he wasn't about to disappoint. Eager to complete his deadly mission he was walking and flying by the end of the month.

As Robin awaited his rescue on that warm afternoon, he gratefully supped from a pop bottle, just as the Baron had wisely refused offers of alcohol preferring to take water when his rescuers arrived at the scene.

Whereas von Richthofen had undergone deep psychological trauma, Robin was at least spared the angst of war and the killing of his fellow man. However, those who knew him best would later say that the crash landing in Germany changed him: causing him rather more concern than he might have admitted to. Also, that the incident aged him somewhat, though he was always quite a youthful looking man.

As he sat on the wrecked fuselage with the reporter and cameraman as company, he could plan a new triplane that would have a much better power unit, be more reliable with something better fitting to the lines of the original cowling.

'Ten years flying that aeroplane,' he told them. 'I've probably done about 500 hours in it. It's been to Germany before.' He wanted them to know this had been a good aeroplane.

'Will you be able to fly it again?' asked the reporter.

'I don't know. We'll have to dismantle it and take it back to England, then one day repair it. But it'll be a long job, I think.'

The reporter asked Robin to do a piece to camera to explain what had happened. Fortunately over the years he'd built up considerable experience in interviews with the broadcast media and so this didn't faze him. Of course, his voice would be faded for an interpreter's once the piece was broadcast on German TV.

'Unfortunately, I had an engine failure and I had to land the aeroplane in this very, very small field. And it's very sad for me because I've been flying it for more than ten years now – so it's very sad that it should end this way,' he explained with stoic understatement.

Von Richthofen himself was treated as a celebrity in Germany, frequently photographed and finding himself in the limelight due to his status as an air ace. A quiet man by nature, he hated such publicity. The last thing he wanted was to be feted at lavish balls and parties hosted by industrialists and generals for the good of the war effort and the benefit of meeting the great and the good.

Robin was asked whether the triplane was fixable?

'Well, obviously we'd have to find out what was wrong with the engine, but we can strip the fuselage completely and re-weld, make new wings, new undercarriage. We've got to find what was wrong with the engine. It's a shame because you can see it was a beautiful aeroplane.'

At least talking about the incident helped him to go over in his mind just what the sequence of events had been.

'Did you notice the oil pressure needle ..?' asked the young reporter.

'No, the engine just started going bang, bang, bang, so I closed the throttle and had to immediately find somewhere to land.'

'How high were you?'

'About 1200 feet – 400 meters? Because we have to fly below the Dusseldorf radar – the Dusseldorf control space. That is 1500 feet, so I was 1200 feet.'

'Did you send an emergency radio message?'

'No, there wasn't time to do that.'

Horst Schwenke of the Hamm balloon club and organiser of the show had heard the news and arrived in his private helicopter landing just meters away from the crashed triplane. He had been looking forward to what would have been a unique display but was relieved that at least Robin had escaped injury.

The end of the beginning: the crash that would lead to the rebuild of the triplane

'Don't you worry,' he assured Robin. 'We'll look after you!'

It was decided that the triplane could be moved that very afternoon. Concerned that the men who arrived with the low loader were going to cut off the wings within millimetres of the fuselage Robin insisted he show them where best to cut as the wing roots would be vital in any rebuilding work.

Calculating the Cost

With the triplane safe and secure in a hangar at Hamm, Robin was able to return home to consider whether the triplane really did have a future. He'd built up such a following as the Red Baron he couldn't envisage quitting now. A German crash investigator for the Deutsche CAA estimated the cost of damage to be around 60,000 marks.

For insurance purposes, Robin estimated the following costs in order for the triplane to be rebuilt and based on the assumption that around 80 % of metal fittings, brackets and pulleys et al would be re-usable.

Guide to estimated cost of repair to Fokker Dr.1 G-BEFR:

Materials spruce, ply, etcetera to make three wings including spars, ribs etc, as well as undercarriage spreader and interplane struts: £4,000

Ceconite covering, tapes, stitching, dope, glue, varnish, paint etc: £1,500

Control cables and replacement of small number of fittings and turn buckles etc: £350

Replacement of Cabane struts and fittings as necessary: £800

Replacement landing gear legs, axle boxes, wheels etc, as required: £2,200

Propeller: £1,250

Fuselage rejigging and crack testing to attachment points repairs as necessary: £1,000

Cowling replacement repair: £400

Plans set: £150

Labour to repair and remake as necessary from firewall back and reassemble to flight test standard: £8,000

Final painting: £1500

Transportation from Germany to UK: £2,500

As far as the insurance assessor was concerned, the triplane was a write-off and so B.C. Flight was paid in full. Unfortunately, it had been under insured.

Despite that, after buying back the salvage, Robin could look forward to bringing it home by road and rebuilding it the way he wanted, but this time Pat wasn't so willing. The crash was a turning point and Pat felt he'd gone as far with the triplane project as he could. He argued that the triplane was not a profit maker and he was spending far too much time away from his business – a motorcycle dealership in Saltash.

The role of wing-walker and support partner was an expensive and demanding one, and the idea of spending another season driving across country to run after the triplane on landing no longer appealed. It wasn't just half a dozen displays any more. Every weekend during the summer was being taken up with display flying and sometimes there would be two or three shows per weekend; it had become a full-time occupation.

Even with the increasing demand from shows, displaying the triplane earned the Bowes-Crawford partnership a mere £3,000 per season after expenses, so by the time they'd spilt the proceeds there was little for either man to walk away with. Now, with the insurance payout split 50-50 between the partnership, it was the ideal time for Pat to cut his losses, but with no hard feelings. It had been a tremendous decade for both of them.

Pat's suspicion that it was the Scarab's troublesome number four piston rod that was at the root of the problem was soon vindicated on stripping the engine. The Warner Scarab would be rebuilt but it wasn't going to be put to service in the triplane again. Robin would put in a zero-hour'd, reconditioned Lycoming 0-360.

Working together with friends Dave Silsbury and Ernie Hoblyn the large shed was put to full-time use again for the first time since Howard's death, while Gordon Clarke volunteered to video the repair work as a documentary. In all, the rebuild was a mammoth project that would eventually cost Robin over £17,500, but he was determined to see it through and re-build the triplane in the way he felt it should have been built in the first place. Now, was the opportunity to really put his signature on the triplane.

There was a considerable amount that could be used again with the exception of the cracked fuselage, which was tubular steel. But what they hoped would be a six-month project, eventually became twenty months. Robin redeemed every insurance policy he could and even claimed from HMRC any outstanding VAT or input tax he might have had owing through his purchase of new cars dating from when he'd set up his car sales business. Whatever he could claw back enabled him to push ahead, but he still needed some steady income, and the work he wanted most of all was flying.

With outstanding events in his diary for the triplane during the remainder of the 1992 season, Robin also needed to complete his commitments on the air show circuit. He approached his old friend Anthony Hutton at North Weald who just happened to have among his varied collection of WW1 aeroplanes, a triplane that was largely available for the season. However, the availability of Anthony's triplane did not always fit in with pre-booked dates and it meant having to renegotiate with organisers'; but on the whole, G-BTYV with its Lycoming engine was a very worthy substitute.

Chocolate bars and Flying Circuses

Business partners Vic Norman and Torquil Norman (unrelated) had purchased from the Farnsworth family a remarkable airfield at Rendcomb, near North Cerney in Gloucestershire. RFC Rendcomb Aerodrome was created as a pilot training airfield in 1914 where, for the duration of the Great War and up until 1920, it had been a temporary home for some 3,000 trainee pilots and ground crews training on the locally manufactured Bristol F2A fighter.

In the early 1990s what made Rendcomb truly remarkable was its state of preservation, having not been called into service for WW2, the officer's mess and other buildings such as the mechanics workshop and squash court-cum-hospital were found as they had been left – derelict, but repairable. There was a lot of preservation work to be done before the site could be used again including the reseeding of the grass airstrip, but ministry notes had survived that specified the particular type of grass seed to be used. The grass seed was still obtainable and had originally been chosen because its grass provided a particular softness on landing.

On ploughing up the airstrip in readiness for reseeding, over 200 Lewis gun cartridge cases were unearthed all of which had been date stamped during the years of the Great War. The old gunnery range was still in use as a dry store for hay and straw.

Vic wanted to have the airfield ready to reopen in time for its 75th anniversary. It would be an ideal occasion for Robin to bring the triplane and when broadcaster HTV came to film the preparations for their *Report West* early evening news programme they made full use of the irony of having the Red Baron's triplane at the very site where his RFC adversaries had trained. Describing the triplane as "a colourful prop", the reporter's commentary continued: "The young pilots of the RFC would turn in their graves to see the Red Baron's triplane parked up at Rendcomb."

Vic was not only in the process of recreating a WW1 aerodrome; he was also setting up a permanent and exclusive base for his now established flying circus – the Aerosuperbatics wing walking team that had been based at nearby Staverton aerodrome using his private collection of four Boeing Stearman biplanes.

Like Robin, Vic was a civilian pilot who had built up a considerable record of display flying. He too, had begun his career racing cars – vintage racecars in his case. As a pilot, Vic had learnt his trade flying types such as the Stampe and Zlin 50 with which he did aerobatic performances at over 800 displays including the Monaco and British Grand Prix.

The two pilots greatly respected one another and when Vic needed a reserve pilot for the fourth Stearman he turned to Robin, which was particularly providential following the early termination of TSW's contract for the Pitts Special.

With sponsorship from Cadburys, Vic Norman's flying circus was to recreate the daring escapades of the barnstormers of the 1920s' and 30s' – particularly the American flying circuses that had put so much emphasis into spectacular wing-walking stunts for public displays and Hollywood film makers.

In those days, pioneering wing walkers such as American pilot Ormer Locklear choreographed the most amazing stunts of derring-do despite there being no safety wires or parachutes. Practices that he had personally developed as a pilot in the army such as the tightening of strut wires or plane-to-plane refuelling could now be used to entertain crowds and bring in good money. For spectacular entertainment, wing walkers would openly dance the Charleston on the upper wing or play tennis unencumbered by wires or safety devices of any kind. When things went wrong in a display, as they often did, the wing walker would fall to their death often in full view of the amassed spectators below. It was an unfortunate part of the spectacle, risking all to bring in the money, but that was what the flying circuses were all about.

The Aerosuperbatics wing walkers however, would incorporate a stringent safety plan for each flight using the latest in harness and wire technology to ensure such accidents could not happen.

Wingwalking (credit Aerosuperbatics Ltd)

The team wasn't unique in respect of wing walking in Britain as the Surrey based Tiger Club and later the Northamptonshire based Barnstormers had performed wing walking throughout the sixties though mostly in straight and level flight with the wing walker securely harnessed onto the top wing of a Tiger Moth or similar biplane.

Strong friendships and connections ...

Robin's encounter with Mike Dentith at a hotel bar in Baldonnel in 1986 was to be a fortuitous meeting that would later change his career. At that time, Mike was flying a close formation duo with John Taylor as leader. They were members of Skyhawks – a Biggin Hill display team flying French built Fournier aircraft performing a truly remarkable aerial ballet routine that they had taken to shows across Britain and the Continent. On this occasion they had flown down from a show in Scotland in the most appalling weather conditions in what were essentially powered gliders, but safely ensconced in the bar it was *they* who were particularly impressed by the lone aviator who'd flown a replica triplane across the Irish Sea no less. They had to admit that they at least – though underpowered – had had good forward visibility through a clear, bubble canopy of the Fournier, whereas the triplane was an open cockpit with nothing more than a tiny Perspex windscreen offering little more than wind deflection and zero visibility.

Strong friendships are borne out of air displays and it wasn't unusual for pilots to find common bonds that united them. Mike Dentith, like Robin, had his roots in the South-east of England. The son of an actor, he was a kind, self-effacing man who also had a background in second-hand car sales and since 1975 had been carving a reputation for himself as a civilian display pilot of vintage aircraft. And whereas Robin's creative interests were to be found in music, Mike's was in theatre having trained as an actor he'd worked in various regional repertory companies. For both of them, performing was second nature.

By 1992, Mike was employed as the Aerosuperbatics team leader and it was through his recommendation to Vic Norman that Robin was approached to join the team – now commonly known by their sponsors name: "The Cadbury Crunchies." Robin would be their reserve fourth pilot.

The role of the reserve (or fourth) pilot was part of the solo act flying the smallest of the Stearmans to shows particularly on the Continent. Indeed, flying solo was just the sort of work Robin preferred and so he readily accepted the offer without a moment's hesitation.

For the Crunchies, Robin brought with him not only considerable experience, but also a reputation for being an extremely safe and careful pilot as well as a meticulous flight planner. He was just what Vic Norman was looking for.

Of course, even the most capable of pilots are not simply

taken on. Regardless of reputation, Robin needed to be tested and cleared on the Stearman and once he was familiar with the machine, he also needed to be checked out with a wing walker onboard the top wing. This would be a very different experience for the man who normally flew as the Red Baron.

His first official assignment was to an air show in Berlin with Rachel Taggart as wing walker. He was the ideal pilot not only in his experience of flying to the Continent but also in his independence. Finding his own way was his speciality, and Germany in particular was by now familiar territory.

Robin was assigned the baby Stearman registered: G-IIIG and known affectionately by the team as "Gig". Producing 220hp as opposed to the larger 450hp of the formation team aircraft, Gig was the more economical of the fleet and thereby better suited to the long distance events. The larger Stearmans limited range of approximately two hours flying time meant fuel stops for more distant destinations, and were a critical part of the flight plan.

Gig was certainly unique in that is was then the only Stearman in the fleet to be fitted with a British registered swivel rig that enabled the wing walker to "swaddle" (swivel) upside down in flight independently of the aeroplane. This was usually Helen Tempest because she was the shortest of the wing walkers. Robin also found that one of Gig's great flexibilities was that it was agile enough to complete a manoeuvre that its larger contemporaries were unable to pull off: a "flick" departs from a conventional roll in that the pilot flies a corkscrew around a horizontal line rather than merely inverting and turning the aeroplane over whilst keeping the nose and tail in the horizontal. The flick requires a slight dive for speed and then a pull up which almost stops the aeroplane in what is quite a violent manoeuvre, but nevertheless one that the wing walkers thrilled to – especially Helen Tempest, the team's most experienced wing walker.

Robin never sprang any surprises on an unsuspecting wing walker. Before each show, he would go to great lengths to take the wing walker aside and explain just how he was going to fly the aeroplane and what to expect. Where the solo act was his concern, he would do as he did with the triplane or the Pitts or Nimrod or any of the aeroplanes he displayed. He'd take time to work out a routine that made the best use of the environment – the crowd-line, tree cover or nearby bluffs; and it was his etiquette to explain in detail the plan to either Helen or Sara or Rachel or whoever was flying with him that weekend. He included them, as he believed that was only fair and decent.

Helen had first wing-walked aged 15 having cajoled her aviator father, Barry Tempest, in to taking her up. In 1963, Barry together with Charles Boddington, was a co-founder of the Barnstormer Flying Circus where wing-walking had made a triumphant but safe return to British skies thanks to the development of safety harnesses and rigs that supported the wing walker.

With the idea of curing his daughter's "mad" obsession, he eventually agreed to take her up on the wing. However, it was from that moment that a seed was sown that would lead to her becoming the UK's best known and most experienced wing walker.

Helen first met Vic Norman in 1986 at the World Aerobatic Championships where he mentioned the possibility of her wing-walking for him. Vic was very impressed with Helen's attitude. She had a reputation for stoically wing-walking in the most adverse conditions imaginable come rain, hail and even heavy winds. On one occasion, she'd even left her sick bed to wing walk despite suffering glandular fever. Over a week later following a full recovery, Vic called her with the offer of a permanent job.

In October 1987, she replaced Lesley Gale who was unable

to complete the last show of the season. By the time Robin joined the team, Helen had established a remarkable reputation for herself – a real trouper for whom the display was paramount.

As with Mike Dentith, Robin had first met Helen on the display circuit some years earlier in 1990, together with her pilot, the legendary Brendan O'Brien. They had just taken part in a record-breaking wing walk performed by Roy Castle for his BBC Record Breakers Show.

A wing walker's eye view of Robin

A year later, together with fellow wing walker Sara Cubitt they would "adopt" Robin at any display they were flying at to make sure that the lone flying "Red Baron" was in good company. His reaction to being swept up and adopted by two diminutive young ladies in tight fitting scarlet leather flying suits was not recorded by him.

June 15th 1992: An Evening of Magic and Nostalgia

Away from the display circuit Robin was busy fending off attempts by anxious friends to pair him up with a suitable partner. Following his separation from Wendy, there had been occasional female friends, but nobody serious enough to either detract from or contribute to his mission.

On the evening of June 15th 1992, there was to be a summer ball at Dunkeswell with dancing to a Glen Miller tribute band in the evening preceded by a small air show during the day in which Robin was going to display the Pitts and a Focke-wulf 190 replica, two thirds the size of the original. The 190 belonged to Peter Logsdon who had never actually flown the aeroplane, as he preferred to see Robin fly it. Needless to say, Robin was only too happy to step into the seat. The triplane was at this time a write-off having crashed in Germany only weeks before.

Robin's friends, Peter and Kay Goody, were going with him but were eager to know if there was someone special he might like to take with him?

'Who are you going to take?' asked Kay.

'Never you mind!' retorted Robin.

'Why don't you ask Sheila Truscott?'

'Never you mind, don't you go interfering. I'll sort myself out, don't you worry.'

And sure enough, he did just that ringing Sheila two days before the dance.

'Are you busy next Monday?' he enquired.

'No.'

'I just wondered if you'd like to come to a dance?'

'What? Yes! I'd love to!'

Sheila didn't have to think twice. She and Robin had always

got on well and in effect this wouldn't be the first social occasion they'd had together as he'd taken her out for a meal in the October following Howard's death; but that had been more to do with winding the business up.

He'd wondered then about broaching the subject of getting together more often and maybe with a view to a more permanent basis as they were such good friends and had so much in common; but for Robin, it just seemed too close to Howard having passed away.

The dance provided the perfect opportunity to pick things up again. Sheila had had to work during the day and so wasn't able to see the flying display, but Robin was only too happy to drive back down and collect her for the evening soiree. The Devon Strut (the local branch of the Popular Flying Association) were all in attendance and by chance a visiting Liberator that had been touring the UK was able to do a flypast in tribute to the many US Navy personnel who had maintained and flown the type from Dunkeswell during the war years.

It was an evening of magic and nostalgia, and Sheila was falling head over heels with the man that she'd admired for so long. When the party was eventually over, they followed Peter and Kay back to their home in Ivybridge for coffee before returning to Ermington where they sat talking in Robin's black Citroen under the archway of the Old Inn House until nearly 5 am.

The evening had been a revelation for Sheila. She'd seen Robin in a new light having always thought of him as the typical bachelor, not needing anyone else and it had never occurred to her that he might be the slightest bit interested in her.

The first time Sheila went to Rendcomb was to meet Robin on his return from a show at Duxford where he was flying as the fourth pilot for the Crunchies. Vic Norman had organised a big barbecue for the team, their families and all his many friends, prior to an air show the following day. She and Robin had been booked into a splendid guest house courtesy of Aerosuperbatics and so it promised to be a fantastic weekend away for both of them. This was a new world for Sheila, who was looking forward to meeting Robin's many aviation friends. By nature she was a great supporter of people and loved nothing better than to get involved with whatever they were doing. Howard had been a fanatical motorcyclist in his spare time and Sheila had accompanied him on many rallies getting to know his friends in the process. Now she was supporting Robin and loving every moment of it.

Saturday evening, just as the barbecue was getting underway, the Crunchies were reported overdue having displayed earlier in the day at Duxford. News got through to base that they were weather bound being unable to get beyond Hertfordshire. They were only an hour away, but what with weather and the approach of darkness this would be an event they would be unable to attend much to the team's consternation.

Vic Norman's wife, Anne, was determined that all her guests should enjoy the barbecue.

'You'll meet all sorts of people,' she told Sheila, urging her to stay.

Indeed, it was an ideal opportunity for Sheila to meet people that she would actually be seeing quite a lot of during the coming years, and without Robin by her side she would be meeting them on her own terms.

It wasn't long before she was being invited to join a large table full of men who were all very curious as to who she was and what she was doing there.

'Are you on your own?'

'Yes.'

'Come and sit with us. We'll make room. Come and sit here.'

'Oh, are you sure?'

'Yes, yes. Who are you?'

So she told them who she was and where she was from and that her boyfriend was the fourth pilot with the Crunchies and that they were unable to get home because of the weather.

'Do you fly?' one of them asked her.

'No, but I'd love to learn and I'd love to be a wing walker, but I'm too heavy.'

'You'll have to be a bit lighter than that, won't you?' was the cheeky reply from a very dashing, moustachioed young man who appeared to be the nominal leader of the group.

Unabashed Sheila asked him: 'Do *you* fly?'

'Yeah, a bit.'

'What do you fly?'

'A Red Arrow!' came the reply and the whole table roared with laughter at her expression. Without suspecting, Sheila had been sat with the entire team and had been duly caught out, as had so many unsuspecting ladies before her. Far from being annoyed, she was thrilled to make their acquaintances and considered the experience her entrée into the aviation world and display flying.

Robin and Sheila

Pulling Out All the Stops:

By the autumn of 1992, once the insurance policies had eventually paid out and he was able to find new lodgers to rent part of the Old Inn house, Robin was able to start rebuilding the triplane in the workshop along with friends Dave Silsbury and Ernie Hoblyn. However, the rebuild meant that he was less able to spend time on the correspondence needed in organising the following season's air shows. Pat Crawford had always contributed to this side of the operation, but without his partnership great piles of paperwork were building up in the office and hardly any of it received Robin's full attention and thus he became frustrated at not being able to sort it out. Only the VAT returns got top priority, with everything else going by the board.

At that time, Sheila would immediately have taken on the role, but she was on holiday touring Australia and the Far East with her younger daughter, Helen. She hadn't wanted to leave Robin, but the family holiday had been arranged following Howard's untimely death and some time before her relationship with Robin had begun. It had been intended that

Sarah go too, but other commitments prevented her from doing so.

As she toured, Sheila kept Robin updated with their travels by sending regular audio tapes describing each day's events. In turn, Robin would send cards and letters to the various hotels Sheila and Helen would be staying at en-route. Sheila couldn't imagine leaving Robin for six weeks especially as their relationship was just getting underway, but with the rebuild he had plenty to be getting on with and there would be little she could do to help in the workshop.

In fact it was a crucial period in the rebuild with the original estimate of six months having to be reconsidered. The project was more complex than had been envisaged and was now into its fourth month; it was becoming clear that the triplane certainly wasn't going to be ready for the coming 1993 season.

After five weeks touring Australia, Sheila and Helen flew from Sydney to Singapore where they had been due to stay for four nights. The hotel was on the edge of the city and to access the tourist areas meant a cab ride every time. The air conditioning had the effect of turning their room into a fridge cooling it to such a degree that they propped open the door to the corridor to let in warm air. When they'd arrived at the hotel, there had been a birthday and Valentine's Day card from Robin, thoughtfully sent ahead of time for their arrival on the 14th of February, which was also Sheila's birthday. It was a lovely surprise, but it was the only thing that went right on the Singapore leg of their tour. In their forays away from the hotel they were unfortunate to encounter something of the seedier side of the district and this was enough to cause Helen to retreat to the hotel room and insist they stay put. Feeling vulnerable, they needed to get home ahead of their scheduled flight.

Unable to get an earlier flight, Sheila was in despair. Helen refused to move from the room and it was only when Robin called to wish her a happy birthday that Sheila mentioned their predicament. From home, Robin went to great lengths to get them on a flight out of Singapore by calling anybody and everybody who just might be able to get them on a flight. After considerable effort two seats were found on a BA flight that would be leaving Singapore that evening. He rang Sheila with the good news and within half an hour, she and Helen packed and were on their way to the airport.

Thirteen hours later they were overjoyed to see Robin waiting for them at Heathrow – even Helen gave him a hug she was so pleased to see him. He'd pulled out all the stops to get them home and even driven overnight in order to meet them.

Forays

Robin was well known among his friends for his quiet ways and generosity of spirit. Only that August, he'd promised to help out Belgian Air Show organiser Gilbert Buckenberghs, following the burglary and theft of takings immediately after the Sanicole Air Show in which Robin had flown "Gig". Gilbert had appealed for help from those aviators who had taken part and Robin was quick to volunteer with the Pitts Special asking no recompense other than fuel and accommodation.

He loved having Sheila join him on forays up country to find parts for the rebuild; and even if it meant leaving home at the crack of dawn to travel hundreds of miles, Sheila was more than happy to be part of it.

After a particularly long trip to Caterham having picked up various parts for the triplane, Robin took the slower, A35 coast road in preference to the motorway – he avoided motorway driving as much as he could preferring the more scenic routes across country even if it did take a little longer. As they approached Charmouth he asked Sheila if she'd ever visited there as it sounded such a nice place.

'No, I've never been. Have you been?' she asked him.

'Er, no, I don't think I have.'

'I'd like to go there one day.'

'Well, we can go there now.'

'Bosey! It's dark now. We'll do it another day when it's a bit lighter and we're not quite so late.'

'You don't have to get home for anything do you?'

'Not really, but you're tired, we've had a long day. You don't feel like going there now, do you?'

'Oh yes. Let's go now. There's no time like the present.'

It was mid summer and well past ten o'clock not yet completely dark, but by the time they parked and gazed out over the sea wall the sun had well and truly set.

'Let's climb up that hill,' urged Robin.

Sheila needed no urging. It was a wacky thing to do, so typical of Robin as Roger Wilkin would have told her. Such things have to be done in the now while the opportunity presents itself. Who knows about tomorrow?

So they clambered up the hill falling into rabbit holes as they went until they were at such a height as to look down upon the warm glow of the town's distant lights. Life didn't need to be planned before doing something like this, because it was the spirit of the moment that mattered.

For half an hour they sat arm in arm talking while watching the bobbing navigation lights of the fishing boats.

'There! That's worth it, isn't it?' he asked.

A different kettle of fish

Sheila's strength lay in the fact that she was more than willing to understand and thoroughly enjoy whatever Robin was doing. Although a very capable man, Robin wasn't a home-maker or house-husband; nor was he a provider. He was, in all respects, a plane maker and flyer and the large shed at Ermington was his natural home. Sheila understood this perfectly from day one as if she'd been looking to share this same ambition her entire life, and it was where – following his divorce – other prospective suitors had failed. Sheila not only loved the flying and the driving, she also loved his drumming in the band, whether it was with the Russ Thomas Show band or Cat's Whisker.

Robin's many friends in both the aviation and music worlds liked her instantly. She fitted in wherever she went, as she was always supportive of Robin and genuinely loved whatever he was involved in.

Friends were becoming familiar with the possibility that when they rang Robin, Sheila would answer on his behalf. She was the new unofficial partner in what had been Bowes-Crawford Flight and she was taking to the role like a duck to water.

What was not so easy for Sheila was watching the occasional problems as they occurred. She was now emotionally involved and well aware of the dangers that faced Robin when flying either his own or someone else's aeroplanes.

The Pitts was not so bad; it could land on either grass or tarmac, but the replicas gave her more concern. With his own triplane still undergoing a rebuild during the 1993 season, he borrowed Anthony Hutton's triplane for the Plymouth Air Show – a long standing booking that he really didn't want to cancel. Sheila was on Plymouth Hoe to watch the display when, on his final flypast, something was seen to fall off the aeroplane landing harmlessly in Plymouth Sound. He'd flown down from Bodmin airfield at Cardinham with the intention of returning to the show with the Pitts Special – parked ready to go for the second part of the display. As he set course for Bodmin, Sheila and Ernie jumped into the car and set off in pursuit radioing through to the ATC at Bodmin that something had come off the triplane that could affect the landing. Could they check as soon as he was overhead?

Racing for Cardinham, Sheila and Ernie knew only too well that the triplane is a fragile machine with a lot of vices especially in respect of its undercarriage, which comprises old fashioned, spoked motorcycle wheels. These are not even angled on the Fokker triplane though many other types of the period were. Small, thick wheels on modern aeroplanes are angled and constructed to counter the effects of side loads. Spoked wheels, however, don't take side loads, which is of particular concern to the pilot if attempting to land in a crosswind. Any gusts were a major problem as a triplane could be blown off course.

A slow flypast revealed that on this occasion it was nothing worse than a cover plate that had fallen from one of the wheels exposing the wire spokes of the motorcycle wheel.

During that season he had quite a few commitments to fill with Anthony's triplane, which had slightly different markings and a more reliable engine in the form of a Lycoming 0-360. For the rebuild, he'd get hold of a similar Lycoming as a replacement engine for the troublesome Warner Scarab.

There was one display at Rendcomb that Robin didn't want to do because he hadn't tested the aircraft prior to the display. It involved another triplane – this one from Duxford was to be displayed prior to a forthcoming auction at Rendcomb.

Vic Norman was keen that it be taken up and put through its aerobatic paces but the engine of this particular machine didn't have an inverted fuel system rendering it unable to perform any manoeuvre that would involve a negative G loading. Against his better judgement Robin took off for an acrobatic display and was in the process of performing a stall turn when the engine cut putting the triplane on its back. He wasn't that high, but in order to restart the engine it was vital that he point the aeroplane at the ground in the hope that it would "windmill" and restart. If it failed, he'd have to land immediately in whatever space was available.

As if by a miracle, the engine fired back into life and he valiantly managed to rev it and pull out narrowly averting a disaster. The spectators, unaware of the drama being fought out above their heads, thought this manoeuvre was fantastic! Among their number were Robin's parents and sister Christine, but Sheila and a friend realised what had just happened and without a word to one another started to run.

Completing the display so as not to cause concern, Robin eventually ended the routine and landed. Sheila, to her credit, normally never expressed any worries she had when Robin flew, but on this occasion, she had to know.

'What on earth happened!? What on earth happened!? How ...?'

'No, no, it's all right, don't worry, don't worry,' he assured her. But between them they knew that he'd come very close to a disaster.

Any publicity is ...

Robin always welcomed publicity surrounding his flying, and over the years the regional press and TV had featured him on various occasions. A keen amateur writer, he'd even supplied his own articles to national aviation magazines such as *Wingspan*.

However following his near fatal dive at Rendcomb, the *Daily Mail* ran a piece that claimed that it was Robin who was selling his triplane.

"*Swooping in to Land – A Legend: Red Baron Replica is up for sale* by Paul Keel: The British used to run for cover when they saw her coming but today she's a crowd puller. This replica of a Fokker triplane flown by Germany's Red Baron Von Richthofen in the First World War was built for the film The Blue Max. Accustomed to celebrity status she upstaged stars George Peppard, Ursula Andress and Jeremy Kemp. She is expected to fetch more than £50,000 in an auction at Rendcomb Air Show in Gloucestershire today. The original triplane's unique design gave it extra lift enabling von Richthofen to shoot down a record 82 allied aircraft.

"The replica built in Germany in 1965 later went to a Florida museum. She was rescued in the mid 1980's by enthusiasts.

"Air Show pilot Robin Bowes has clocked up 500 flying hours – more than the Red Baron. 'But,' he says, 'the plane remains a challenge. Flying her is a bit like standing on one leg. You have to concentrate all the time.'"

Test flying the Duchess

When not testing his own aeroplanes, Robin could be found acting as test pilot for fellow aviators. Here he poses in front of Nigel Hamlin-Wright's glorious Avro 504K

Robin's knowledge of, and interest in, biplanes and triplanes introduced him to like-minded souls across the world. Some he got to know well whilst others remained mere correspondents. Of the former, one such comrade of the skies was Nigel Hamlin-Wright who had grown up in various RAF bases in the UK and Germany, his father having been a serving pilot. When Robin met him in the early 90s, Nigel was ensconced in a remarkable project: the rebuilding and testing of an Avro 504k – much loved in its heyday by RFC pilots and later flying circus barnstormers such as the great Sir Alan Cobham.

Nigel documented this process in his book *A Standard Pilot's Notes* and published in 1996 by Chocks Away. Rather than reworking Nigel's excellent records for this book, the author has requested that he might include passages here in testament to Robin's collaboration on the Avro project.

"The next one I saw [Pietenpol Aircamper, a two seat, open cockpit parasol affair, designed in 1929] was built by a chap called Dave Silsbury. From his picture in a magazine he looked an affable chap so I gave him a call. I was right. He was charming and straight away agreed for me to come down to Dunkeswell where he kept his pride and joy. What's more, if the weather was right, he promised we'd go for a spin."

"Rolling into Dunkeswell airfield about mid-morning, I was directed to Dave Silsbury sitting in the café and went over to introduce myself. Another chap was with him, sitting in flying overalls and sipping coffee: Robin Bowes. They both had an easy manner, didn't appear to take themselves too seriously and were certainly happy to include me in their jokes. I took to them without hesitation.

"Half an hour later Dave and I lifted off in the Pietenpol and Robin joined us in his Isaacs Fury though well throttled back. The whole scene was thoroughly captivating. Here was I flying in formation in an open cockpit vintage aeroplane, both new experiences, and in the company of people whom I felt were to become firm friends despite our brief acquaintance. I saw that my life was momentarily charmed.

"Getting to know these two men had a greater significance in the scheme of things than I could have imagined at the time for I was to meet Robin a year or two later at Dunkeswell to discuss with him my ideas for the Avro.

"'Any help you need old mate, give us a call, I'll see what I can do.' This was Robin (and Dave) all over, and when I needed an engine mount for a Scarab, it was to them I turned.

"A month or two before I called him, the engine in Robin's Fokker Triplane had failed over Germany. The subsequent hurried landing had seriously damaged and almost written-off the aircraft. It provided him with the opportunity to rebuild and re-engine the Fokker with a flat-four Lycoming. Consequently, his Scarab and its mount were hanging around waiting to be fitted to a Sopwith Triplane.

"After a brief explanation over the phone, Robin brought the mount up to AJD, which gave him the chance to check on the Avro's progress."

"As my thoughts turned to the practicalities of operations on the Air Show circuit, my hand instinctively reached for the phone. I called Robin. He had been plying his trade up and down the country disguised as the Red Baron for well over ten years and there were very few surprises left for him.

"Robin and his immaculate red Fokker had become over the years entirely self-sufficient, save for the business of fuel. In his aircraft and hidden about his person, he carried all the things that make life easy for when arriving at an air show. Chocks, a clever device that locked the controls, a miniature tail-dolly, locking wire, assorted nuts and bolts and a modest set of tools were strapped down, clipped up and tucked away in corners and under the seat.

"We discussed things like the desirability of brakes, a battery for the starter motor, various little bags and clips to stow and secure personal effects and, most importantly, a tail-dolly for manoeuvring the aircraft in and out of hangars and across miles of concrete. The undercarriage was now too far advanced, in fact it was complete, for the addition of brakes; there was nowhere in the fuselage that we could stow a battery without a major modification; bags and clips I could deal with in the fullness of time, as well as the tail dolly."

"Keeping my hand in, and hoping to meet Robin, I clattered off to Old Warden in the Cub for a flying evening and a bit of face time. … There is nothing to compare with flying to an air show except flying to an air show at Old Warden, and if you can manage it in an aircraft of at least some vintage then so much the better. The welcome is warm and the ground crew unstinting in their efforts to accommodate you.

"I closed the throttle on the down wind leg and was pleased to judge my arrivals to a nicety (there is a bumpy bit in '22' about fifty yards in). Robin was there in his triplane."

"As the last act landed, I followed Robin out onto the runway for an immediate departure. Half an hour later I was abeam Cambridge …"

"She was as pretty as a picture at the end and looked absolutely 'right'. I had no doubts that she would also 'fly right' come the test-flight.

"There followed some argument as to who would have this privilege. I was informed that a CAA test pilot would have the job. I was not happy. I had decided a long time ago that Robin would be the man and had asked him if he would oblige. My reasoning was that Robin continually flew aircraft of this vintage and, what's more, was thoroughly conversant with the Scarab. I had seen him fly, watched him as he did his walk-rounds, and liked what I saw. He was no 'kick the tyres, light the fires' merchant. I put my case to the CAA and, thankfully they acquiesced. Bloody Hell! It was my aeroplane after all.

"But it was sometime before this happy event took place. Nor had we decided where the test flight would take place. Cambridge, Duxford, Old Warden would be nice. I left it for Tony to arrange. Robin, need I say, was delighted to accept the job."

"Insurance had been a bit of a thorn in my side. I had been quoted quite ridiculous sums by 'specialists in vintage aircraft' that would have made the whole project untenable had I enquired at the outset. It wasn't until Robin came to the rescue, giving me the name of his insurers, that I got a sensible figure with which to negotiate."

"The weather on 30th September 1994 was not bad: a bit overcast, a little wind and a slight chill in the air. Almost perfect in fact. I had asked about 30 people to come along for the event and Marshall were very good about it. At first they weren't too keen on the general public wandering around bumping into top-secret military projects and filling their pockets with souvenirs. I gave assurances that my friends weren't like that and everyone would be exceptionally well behaved. What's more, I would personally see to it that all their passes were returned to reception on the way out. They agreed, if a little nervously.

"The final inspections complete, we pushed the Avro from the hangar and set it down on the grass in front of the tower. Tony and I exchanged documents, which included the final payment, and fuel and oil were topped up. I had completed engine runs and checked all the systems the week before, so all that remained was the business of photographs and the wait for Robin who was driving up from Devon to arrive at about lunch time.

"While having lunch in the canteen at Marshall, I was surprised to see that there hung on the walls a small collection of lithographs by one Edwin Ladell. Printed in about the early fifties they represented scenes of the more famous parts of Cambridge and I wondered how they came to be there. Joan Ladell, his wife, has long been a family friend and has painted most of my family over the last 20 years. It was comforting, in a funny sort of way, that Edwin, who died before I had the chance to meet him, was here taking part in the proceedings. Mrs Kettlewell, a great one for signs and portents, would have approved.

"Robin was later than expected and it was nearly three o'clock when he climbed aboard and I swung the propeller into life. After the run-up I moved to the side of the cockpit and wished Robin 'Good Luck'. I had no reservations about the successful outcome of the first flight, but I didn't envy him his performing in front of an audience. He smiled the smile of a man preoccupied and motioned for me to remove the chocks.

"The wind on the day dictated that we used the grass runway '23', which was a long way from Hangar Two, the tower and our party of well wrapped well-wishers. Tony took one wing and I the other and we jogged with the Avro for almost quarter of a mile before swinging her into wind, ready for takeoff. A final 'Temperature and Pressure OK,' 'Straps Secure,' 'Full and Free,' a wave, and the throttle was opened.

"The magic of this moment as the Avro left the ground, in under thirty yards, was not lost on our assembled company. The majesty of her transition from earth to her natural state was underlined by a curious levitation in an almost horizontal attitude. There was no undignified clawing for height with her nose thrusting heavenward, she simply rose, unhurried, confident, magnificent. She bewitched us all.

"Moved by this stirring sight, Tony and I shook hands. The last two years of struggle melted away as we withdrew into our private thoughts: Tony justifiably proud of his first reproduction aircraft to fly in England, and I recalled the extraordinary nature of my own transition from plongeur to owner of a 504 in what suddenly seemed two short years.

"There was little time for further reflection as the Avro adopted a different note, beginning to 'blip' like a rotary. There was obviously a problem and my heart sank. Was the engine going to prove to be a white elephant? A dozen costly scenarios flicked through my mind while Robin struggled around a shortened circuit, losing height rapidly as he curved in towards the field. My heart levitated a lot less gracefully from my boots to my mouth before Robin executed the gentlest touch down I had ever witnessed and rumbled along the grass, the propeller windmilling to a halt. Tony and I ran up to consult.

"Ever the professional, and speaking as though nothing had happened, Robin had diagnosed water in the fuel. I prayed that he was right. And so he was. From the tank and the carburettor sump we drained nearly a pint of water. This struck me as odd. I had drained the gascolator before the flight and no water was present. We could only guess that it must have collected in the back of the tank as the aircraft sat in the hangar over the last few months, and got into the system when the aircraft was in the flying position.

"We refuelled and Robin took off again into the setting sun to put a seal on the day's success. I went back to the hangar and thanked everyone for coming along. It had been an exciting day and I was beginning to flag from its mental and physical demands.

"Robin, his girlfriend Sheila, myself and Sue, closed the hangar doors and went into Cambridge to Brown's Restaurant for a well earned slap-up. It wouldn't take much imagination to guess at the subject of conversation."

"The following day we returned to Cambridge and the Avro. I had my first trip. Aside from the exhilaration of flying in an Avro, something which not many people in the last sixty years have been privileged to do, I was aware of being piloted by someone set apart from the ordinary. It wasn't Robin's precision that struck me for I had no instruments in the front to measure it, but his lightness and fluidity with the controls. I know it's an old cliché, but if there are people for whom an aeroplane becomes an extension of their limbs and spirit, then Robin ranked among them.

"He signalled to me, stirring the stick, to take control. I felt like a clumsy pupil and restricted myself to a gentle turn or two for fear of being thought of as incompetent. Not many minutes passed before I handed back.

"My friend Kevin and I had fitted a small hand-held radio with an intercom. Needless to say, like most aircraft radios of my experience, this proved to be hopelessly inadequate (no fault of Kevin's) and would have to be sorted out before I joined the air show circuit. It was also evident that an alteration to the tail plane incidence was needed as the Avro only flew level with the stick well forward. Two or three minutes of this was quite tiring.

"When we landed we were met by some friends of Robin and Sheila's; Joseph Koch and his devilishly charming wife, Brigitta. Over from Germany, where Joseph had collected a number of vintage aircraft, Robin had alerted them to the test flight and they had made time to call in at Cambridge.

"Robin took off again, this time with Joseph in the front. Ten minutes later Joseph climbed out beaming. He was a man of few words, but those he spoke he meant, and I valued his praise for the machine. In the absence of Mrs K, I made a mental note to pass on the Koch's enthusiasms. To round off the day, we had an excellent meal at the Butterfly Hotel in Bury St Edmunds. The following morning I woke with a feeling that the events of the last few days might well have been a dream.

"From the moment Robin and Sheila left the weather turned against me, and it was six weeks before he returned to continue the test-flying programme. When he did, we drove to Cambridge early one morning and decided that it was time to get the Avro home. After a quick circuit, Robin landed and we pushed her a good 200 yards uphill to the pumps to refuel. I was reminded of our earlier conversations and, my arms about to collapse under the weight of the tail, I resolved to sort out a tail-dolly as soon as possible. We took off, and after a farewell pass in front of the tower, set course for the farm.

"As I knew the area I was given the stick and Robin sat back to monitor the engine instruments. We were due to be in the air long enough to use three quarters of the fuel in our tank, an endurance that was called for on the flight test schedule.

"I scanned the horizon in roughly the right direction, searching for the chimney and silos of the British Sugar factory about thirty miles away. As soon as I saw it I knew I could relax and enjoy the trip.

"Having picked it out fairly quickly, I settled down to wonder who might be down below; perhaps some one-time joy-rider looking up and not believing his eyes, or some aeroplane spotter unable to make out what this strange craft was. Then I wondered if Lakenheath had us on their radar and if they were puzzling over our lack of progress.

"Over the farm I pointed down to my triangle of grass, which Robin acknowledged, and I handed back control. I faintly heard him shouting if I wanted to try a loop. I suppose he was disappointed in my reply but he didn't insist. He opted for stalls and a spin or two instead. He hadn't mentioned a choice!

"The passenger seat in the Avro is exactly on the centre of gravity which ensured I had no ill-effects as my horizon toppled and turned about the fuel filler cap in front of me. I even had time and the presence of mind to take some photos.

"An hour and a half after setting off from Cambridge we landed at the farm for the first time. The grass was about six inches high and the landing roll was noticeably short. The approach was very steep by normal standards, and at first I thought we were going for a low pass. But when the throttle was shut, the Avro was practically stopped in mid-air; such was the drag from all the struts, wires and undercarriage.

"The second phase of the operation was complete and I felt much happier now that the Avro was home. The test-flying could be attended to more easily and I could see to the business of cleaning and inspecting with ease. Robin left for Devon once again with the thought that his next visit would see the end of the test programme.

"And so it was. We tried a couple of loops this time. The first was a bit slow over the top and I got a face full of oil and fuel from the vents in the filler caps. The second was perfect. It was almost enjoyable."

"I called the CAA to tell them that all was well, the report was in the post and I looked forward to receiving the Permit to Fly.

'Not so fast,' they said. 'Our man has got to come and fly it first'.

'Why?' I naively enquired.

'Because we say so,' they replied, 'and you won't get the permit until it's done.'"

"… I spoke to Bob Cole on the phone and arranged a day for his flight. He duly arrived and the weather turned as he knocked on the back door. There was nothing for it but for him to stay over as the Met Office promised a fair morning the next day.

"I had never entertained a test pilot before and was intrigued to know what manner of man he was and what he had flown. Just about everything, it seemed, from VP1's to Tornado's. I could only assume that he was up to speed as it were. … Bob had indicated that I was to accompany him on the test flight, but I wasn't particularly looking forward to it.

"We were up fairly early the next day and as promised it was bright and perfectly flyable. A couple of friends from the village had volunteered to come and do some pushing and shoving as the ground in front of the 'agricultural building' was a bit on the soft side. I had been stuck there before, almost up to the axle in oggin.

"The flight lasted for forty minutes and we went through the same routine as I had with Robin. There was a difference. Bob was testing the aeroplane with much more aggression than Robin and I was far less comfortable. This is not to cast any aspersions on Bob's ability as a pilot, which was never in question from the moment we took off. You know when someone's 'got it' and Bob certainly had.

"I was pleased to see that the Avro rolled level from stalls in the turn, both power on and off. Now that I was paying more attention, I was happy to see that height loss after a couple of turns of spin, both left and right, was minimal. I knew Bob wouldn't waste the energy at the end of the dive to Vne (Velocity, never exceeded) and we pulled up and over into a loop to complete the exercise.

"We came back to the field, having conducted the test over Honnington, and Bob set us up for the landing. I wasn't certain that his method of leaving the throttle slightly open was going to work. My suspicions were confirmed when he initiated a go-round. He set us up again and repeated the procedure. Maybe it was part of the test. The third time, just over the hedge, the engine died as we touched down, then roared, fell silent and roared again. I knew something was wrong. We taxied back to the top of the field in this mode. I guessed he was throttling with the switches. Shutting down, Bob shouted to me that the throttle was stuck open at about 1000 rpm. I was glad I had a test pilot on board.

"Now all we needed was the final inspection by Terry Newman, and the Avro would be finished. My thoughts turned to the launch party that Mrs K and I had fixed for Easter Monday. And there was the business of the throttle to attend to."

"A last minute call to Robin confirmed that he would be coming up by air, although Sheila was unwell and couldn't attend. Easter Sunday was bright with not much wind; Easter Monday dawned cold, grey and blustery. Typical.

Nevertheless, several of my flying chums braved it and flew in. The line of parked cars and aeroplanes stretching down the strip made a splendid sight. Robin and Dave Silsbury flew up from Dunkeswell in Dave's Jodel and the 'Aerial Demonstration', as advertised, began.

"It was a very jolly event. Sally and Fraser's children ran the bar, Belinda ran the carpark (hence the mix of cars and aeroplanes) and Vicki and Sally ladled the soup.

"Toward's the end, after Robin's display, we drew the number for the prize of a ride in the Avro. It went to Mrs K's close friend and mentor, Paul Lambillion. …

"I noticed when they were coming in to land that the throttle had stuck open again. Luckily, no one realised there was a problem and Robin coped with his customary professionalism."

"Robin had introduced me to Dennis Neville who would act as display pilot when Robin's schedule was too full (most of the time it seemed) and Dennis put his finger on the problem immediately. The gap between the tops of the cylinders and the cowling was far too big. In fact, it shouldn't have been there at all. The air was taking the easy route, which was over the tops of the cylinders, rather than through the cooling fins. The gap had to be blanked off."

"By now the weather was gradually getting warmer and my first booking with the Avro was imminent. North Weald was to be its public debut and as Robin was flying his Triplane, Dennis was to be the ferry and display pilot. (I had not yet solo'd, in fact, I hadn't even flown from the rear cockpit at all.) We were to arrive at North Weald on Friday evening and stay the weekend in a nearby hotel with the rest of the aircrew. It promised to be an exciting event."

"To check on the cylinder head temperature, we decided that Dennis should take her up for 10 mins before we departed. But it was not to be. Dennis landed and 'scratched the show' as they say. The temperature was over the red line and he wasn't prepared to risk it. Maybe if Saturday was cold enough we would try again. We put her away, both of us deeply disappointed.

"As we closed the hangar doors a little silver Stampe buzzed over and came in to land. It was Robin and Dave Starkey (Dave was also on my list of display pilots) who had come up from North Weald to see what had happened to us. We all stood in the hangar and discussed the problem. It was clear that some quite serious remedial work had to be done on the cowling. However, we agreed to stick to our plan and give it another go in the morning."

" ...Robin had called to say that we had a late booking at Southend the following week. Things were beginning to slot into place and the excitement of the season was on. There was still, though, the niggling problem of the oil temperature."

"After putting her away, we were whisked off to an hotel and I made for the bar and a well earned sharpener. By degrees, a few aircrew filtered their way in and I met for the first time people I would be seeing again, up and down the country, during the coming months: Doug Gregory and Des Biggs, the SE5 duo; Vic Norman and the Crunchie team; a smattering of acrobatic pilots and display teams from the Continent."

"Everyone was very complimentary about the Avro, though I was more than happy for Dennis to field all the questions since I was a stranger among them. Robin was also there with his Triplane and we raised our glasses to the Avro's first real show. She was going to start earning her keep. Southend would be a gentle introduction.

"It would be an unusual show in that the actual display would be held away from the airfield over the sea front. Seeing the Avro, the Triplane and the SE5s off from the field prevented me from watching the 504s first display!

"The wind kept up over the weekend and the rain came with

it reminding me of another job I had to do – make up cockpit covers. Trish and I got soaked and cold doing our wing-walking duties, but it didn't matter. This was a new and absorbing way of life and, despite the waiting around and incessant cups of coffee in the smoky crew room, the time raced by.

"From the airfield we had an occasional glimpse of the SE5s, the Avro and the Triplane, chasing up and down the sea front. They were all back within twenty minutes of leaving, their pilot's eager to get out of the cold."

"The two days [at Woodford Air Show] went by very quickly and it seemed that no sooner had we got there, then it was just time to leave. At about three o'clock on Saturday, just after our show slot, Dave and Karen set off in the Avro to Rendcomb. Vic Norman had asked, through Robin, if the Avro would attend a party he was giving on Sunday. I agreed, knowing there was a collection of vintage aircraft there, among which was a Comper Swift, a type I had never seen before. I would fly home on the Monday after a corporate 'do' for a local radio station.

"It took me hours to get out of Manchester in the car and I sped down the motorway to arrive at Rendcomb at seven o'clock in the evening."

"Robin and Sheila took me to a local pub where I was to stay for the next couple of nights."

"Monday's weather was not so kind. … Black clouds glowered to the east and prevented my departure in the afternoon. That night, at the bar, I learnt about the foibles of the local weather, but was assured of, almost guaranteed brighter things on the morrow. Nodding heads and knowing looks confirmed this collective opinion. I had borrowed £10 from Robin and stood my round. Of course, they were absolutely right. Tuesday morning dawned misty, but a hint of my companions promised sunshine signalled an improvement by lunch-time."

"... I laid the map on the hangar floor a dozen times, studying it in the hope that the route might etch itself on my mind. Robin could see that I was working hard at this one and offered to fly back with me. I would return him to Rendcomb in the Jodel. I was nearly tempted but knew that I just had to get on and do it."

"... Mrs. K was pleased with the whole day. She was getting to know people at Shuttleworth and since it was only a one-day affair, she decided that this was the show that she would attend in the future. She had first visited the Shuttleworth Collection in the 1970's and a picture of Robin's Triplane in the hangar was amongst her snaps. I don't suppose she had any idea then what would transpire 20 years later.

A Brief History of Flying Circuses and Barnstorming in the Westcountry

Robin operated as a display pilot in an age constrained by the restrictions of controlled air space and complex laws regarding civilian flying. His pre-war predecessors had been more fortunate in their freedoms, but by the outbreak of WW2, the free and easy days were over for the barnstormers.

Robin was the natural heir of flyers like Sir Alan Cobham and particularly of a Cornish pilot, Captain Percy Phillips – the co-founder of the famous Cornwall Aviation Company Limited.

A St Austell man, Phillips had been demobbed from the RAF in 1919 and returned to his hometown where his father owned a cooperage. It was a time immediately after the savage hostilities of the Kaiser's war – as people knew it then – when the newly formed Royal Air Force was selling off its latest aeroplanes including the excellent SE5a and Avro 504K, and they were being sold off cheaply – extremely cheap in fact.

Employed in his father's company, Phillips or "PP" as he was affectionately known, joined the RAF Reserve to maintain his flying skills but didn't consider that there just maybe a living to be had flying as a barnstormer; that is until 1924 when an Avro 504K had been repaired in his father's garage. The overhaul provided **PP** with an insight into the burgeoning world of civilian aviation and how ordinary people were paying to experience a flight over their neighbourhood. The budding young entrepreneur could see an opportunity opening up before him; he could actually use his wartime skills and earn a living through peacetime flying.

It might seem hard to believe today, as civil and military aircraft fill the skies above our homes, we barely tilt our heads to look up, but for our grandparents and great grandparents, it was considered by some in authority that the populace should become "air-minded"; and therefore, display flying that included the flying of ordinary folk for a joyride in the blue skies was firmly encouraged by the authorities as a marvel of the modern age. In total, 150 towns across the UK and Eire were earmarked to receive visits from the flying circuses where people could observe and pay, if they so wished, three shillings and sixpence to experience the marvels of flying.

Between 1924 and 1939, flying circuses such as The Cornwall Aviation Co. Ltd, toured the country providing rides and the most spectacular entertainment for people from all walks of life. Local dignitaries were often the first in the queue – town aldermen, mayors, squires, magistrates and chief constables thrilled to seeing their local area from a bird's perspective; but flying straight and level soon became old hat and punters looked for a more thrilling ride of loops and rolls, which the likes of **PP** were only too happy to provide.

There was the occasional mishap, but generally of the many hundreds of thousands who took to the skies during that period, there remained a remarkable safety record that ensured the continued success of the circuses.

PP employed pilots and engineers to maintain his enterprise, which he took across the UK every summer. On landing and setting up in a suitable venue, so quick was the turnaround in passengers, that the ground crew were worked off their feet helping people to climb in and out; and some passengers were far less able than others to climb aboard, but rarely was anybody turned away through inability to climb aboard an SE5a or Avro 504 as favoured by Phillips and the C.Av.Co.

In just eight years of the C.Av.Co, Phillips introduced some 60,000 punters to the thrills of open cockpit flying and he was soon recognised officially for holding the record in passenger flights.

For many years, raising finance had been a problem until the large manufacturing companies were persuaded that they could reach thousands by having their logos and slogans painted on the underside of wings and along the fuselage. Fitting the Avro with a hook, PP would tow advertising banners prior to a show; and where money wasn't exchanged then products were provided free of charge or at reduced cost such as engine oil or dope for the wings courtesy of any number of eager suppliers. For the crowds, there were souvenirs that could be bought from the various marquees that were erected for the show – making it a true circus. Ground crew sold photographs of the aeroplanes and pilots, metal badges and even models of the aeroplanes. Programmes funded by advertising were sold, too – a job that Capt. Phillips young sons enjoyed in the holidays as they accompanied their father and his team around the various shows.

Capt. Percy Phillips flies two eager passengers for the thrill of it as part of a Flying Circus visit during the inter-war years

Advertising helped to bring in much needed revenue and this also drew more interest from the press who always attended a show whether large or small. Critics were at a loss to find negative comments and the *South Wales Echo* described PP as "one of the most attractive personalities", and titled him: "Loop King".

His abilities in aerobatics marked him out as the daring pilot who stole the show with his amazing aerial performances. Wing walking doubled the excitement and fellow pilots such as engineer Martin Hearn would sometimes accompany him to wing walk or simply sit on the leading edge of the wing without harness, belts or even a rope to hold on to. Seen from the ground, this was nerve-racking stuff especially as the daring stuntman not only clambered about on top of the aeroplane, but would also climb down to the undercarriage whereupon he would tap on the fuselage to signal Phillips that he was ready to loop. Commentators on the ground would entice the crowd into believing that PP was deliberately trying to shake his passenger off the landing gear or else he'd be in trouble.

Nothing could have been further from the truth, of course, as PP was an extremely popular man with the team and the crowds. He was a naturally considerate employer and a calculating pilot who planned his routines meticulously. However, there was one near error involving Alan Cobham no less. PP landed in the path of Cobham's approach and Cobham was furious. On landing, he ran over to Phillip's aeroplane to remonstrate most strongly, but the C.Av.Co. team seeing this as unjust ran to the aid of their leader and demanded that Cobham withdraw his criticism.

Despite advertising and the loyal following of his fellow pilots, support staff and even family members, the Receiver was appointed in 1936 to announce voluntary liquidation for the Cornwall Aviation Co. Ltd, with losses being attributed to that season's bad weather that had begun at Easter and continued through the summer with frequent days of nothing but wind and rain.

Determined as ever, PP launched a new company – Air Publicity Ltd and, having paid off the company's debts, he even bought two Airspeed Ferries for £2,250. The old circus aeroplanes such as the Avro and SE5a tandem twin seaters that had proved such a thrill with crowds were now becoming redundant as their larger successors could carry multiple passengers and at greater speeds. Proper airfields with purpose built runways were being established across the country and domestic airlines were setting up for business with twin-engine aeroplanes such as the de Havilland Rapide.

Yet the Avro wasn't to die off overnight, as PP entered the Devon Air Race flying between three airports – Roborough, Haldon and Exeter – flying an Avro 504N with his wife as passenger, he clocked 103 mph and won the race despite the scepticism of the pundits. Even *The Aeroplane* magazine described the now twenty-year old Avro as a "genuine antique".

The late thirties saw an escalation in pilot deaths through air accidents with some of PP's closest and most accomplished friends losing their lives. The era of the great flying circuses was drawing to a natural close, yet the aim to make the public "aviation aware" had undoubtedly worked. Technology in the form of faster aeroplanes with enclosed cockpits and passenger cabins together with the advent of new laws and restrictions' governing their use was proving a death knell for the pioneers – the aviation veterans of the Kaiser's war.

Fate was waiting too, for Capt. Percy Phillips. In the summer of 1938 he had visited an old friend Mrs Crossley for lunch, flying in with the Avro 504N (G-ACRE) that he had so often used for the circus. Following an excellent afternoon in the company of old friends he thanked his hosts for a splendid lunch, climbed aboard the Avro, strapped his faithful dog into the rear seat and waved a fond farewell.

Ever the showman, he deliberately held back on lift off for a spectacular near miss of the treetops at the far end of the field. Unfortunately, a down draught prevented his clearing the trees and the Crossleys together with a visitor, Capt. Fair, were the only witnesses to the sound of the Avro crashing out of sight on the other side of the trees.

They ran to assist removing him from the blazing wreckage and an ambulance carried him unconscious to Bedford hospital. PP had sustained multiple injuries braking limbs and fracturing his skull.

Mrs Crossley pleaded with the doctors to save her friend's life, but there was little they could do – his injuries were too severe. PP died before reaching the operating theatre; he never regained consciousness.

Captain Percy Phillips was remembered for many years after the end of the war, but as memories faded with the generation who thrilled to those pre-war exhibitions of barnstorming, so too did his reputation as such an excellent flyer and entrepreneur. He is buried in St Mewan, near St. Austell; an RAF crest is the only part of his epitaph that links him to aviation. There are probably many today who have no knowledge of the once great pilot who rests there.

Many of his old colleagues who comprised the C.Av.Co went on to other jobs in aviation during the war years, some as flyers others as technical and ground crew. Mrs. Crossley, who'd witnessed his crash that fateful afternoon, ferried aeroplanes of all types for the Air Transport Auxiliary along with many other intrepid female pilots.

PP would doubtless be pleased to know that from the 1960s, there was a resurgence in the spirit of the barnstorming flying circuses with both original biplanes and their replicas playing a significant role in entertaining vast crowds across the UK. To this day, air displays have the biggest spectator numbers of any national event second only to football. He would be pleased to know too, that it was the pioneering work of companies such as his own Cornwall Aviation Co. Ltd that enlightened the masses to the thrills of flying during the twenties and thirties. Phillips alone had recorded a total of 91,250 passengers that he had personally taken up in his Avro 504N during those years and many more would-be pilots experienced the Avro as a flight trainer once war had begun.

One Avro that belonged to him G-ADEV was restored by the Avro company after the war and went on to star in the film *Reach For the Sky* starring Kenneth More in the role of Douglas Bader – whose real life crash whilst training, cost him the use of both his legs.

C.G. Grey, editor of *The Aeroplane* magazine, wrote this in tribute to Captain Percy Phillips:

"He was one of the kindest and one of the most considerate of men, and all who worked with him loved him. He did a vast amount of good for many and he did nobody any harm – which is the best that any of us can hope to have said about us."

Facing the Crossroads

They were lovely days with much to do. The triplane was nearing completion and would soon be ready. Gordon Clarke had recorded the rebuild on video with Sheila agreeing to narrate the voice over. Once the large shed was clear they could start on the Sopwith triplane – that would be the next project.

Sheila was using her word processor to write to air show organisers and she showed him the benefits of such a system: of how, with the click of a mouse, you could cut and paste replacing whole paragraphs or even whole pages of text if you needed to.

'I'm only an old triplane pilot. I don't understand all that!' he'd tell her.

It was a time for leisure: music was a shared interest, although they varied somewhat when it came to the period and artist. Robin was somewhat old fashioned in his tastes, preferring the music of his childhood – the big bands and concert orchestras of the 40s and 50s. Sheila, younger than him by just four years, loved the more contemporary sounds but there was common ground to be found in a Barry Manilow concert, which they both loved, though Robin never showed his appreciation through applause. And yet he appreciated audience applause when drumming with Cat's Whisker or the Russ Thomas showband.

'Why don't you clap?' Sheila would ask him.

'I wasn't aware that I didn't,' he'd reply. 'I did enjoy it though.'

There was one particular song by Manilow – *The Travelling Song* – that Sheila liked particularly.

'How did he get that effect with the drums at the beginning of the song?' she asked.

'Oh, that's easy! I'll show you next time we're set up.'

'How do you know what to play if you don't read music?'

'Russ gives me a piece of paper with "bang – bang – bang, bang, bang – Bang!" written on it. I simply follow his instructions.'

It was also a good time to introduce Sheila to the family – Harry and Violet, Christine, Keith and Andrew, all of whom lived in Bradford-on-Avon. Robin and Sheila would load their bicycles onto the back of the car when they visited the family, as Bath was close enough to Bradford-on-Avon to be reached by cycle along the canal towpath.

Cycling was such a nice way for them to unwind at weekends. The Camel trail between Wadebridge and Padstow was a favourite because it meant they could relax on the quayside at Padstow and buy a little bottle of wine from the off license and with some plastic glasses simply indulge in one another's company with not an aeroplane in sight.

From Padstow, they could hop aboard the ferry to Rock to see John Betjeman's grave.

Another weekend might find them cycling up on Dartmoor with Peter Goody and his young son. Peter and Robin were close friends and it was a friendship not based on aviation for once, but a friendship purely foundered on shared values and a shared sense of humour. Like a comedy double-act they could make light of just about anything they encountered. This too, had its dangers, as either could fall off their machines at any time during the proceedings through manic hysterics.

Sheila knew just how much Robin needed this and she was determined that he get as much time away from the big shed as possible. He didn't relax easily and when he did eventually sink back into a chair he could fall into a deep sleep he was so tired.

Night times would find them both reading – they loved

books especially John Betjeman's poems.

Leaning against his shoulder, she'd ask: 'Can you see us doing this when we're eighty?'

'Oh, yes, of course.'

But in her mind, she couldn't visualise it.

Robin [on Dartmoor]

Testing

"The C of G (Centre of Gravity) range seemed to be better than on the borrowed triplane, but a direct comparison was difficult." *Robin Bowes*

At the end of January 1994, with the triplane complete, it was necessary for Robin to seek permission from the Popular Flying Association to test fly it. The machine was likely to be quite different from the Warner Scarab powered aeroplane that had let him down in Germany. He could also make comparisons now with Anthony Hutton's triplane, which he'd been flying a lot during the rebuild period.

It was also necessary to obtain permissions that allowed him to fly and show the tripe displaying only its historic German registered number: 425/17 and not its official British registration of G-BEFR. To their credit, the CAA gave full permission.

In the spring of 1995, with the rebuild complete, the triplane was transported by road from Ermington to Rendcomb, where it would be re-assembled for a test flight. A whole day was spent refitting the various component parts and not everything seemed to fit correctly at first, but with a bit of head scratching and jigging about, it all fell into place. The refitting was proving to be an endurance test in its own right with work continuing well into the evening; their wasn't even time to stop for lunch and so any sustenance needed to be taken was "on the hoof".

Early the next morning and despite sunny skies, a force 10 gale was threatening to keep the triplane firmly on the ground. Unless the strong winds abated, Robin would miss the time allocated by the CAA for the test resulting in a further application, which could jeopardise bookings for the entire 1995 season.

Robin – eager for the off – could delay no longer. No force of nature was going to stop him from getting the triplane back into the sky – it had been a long rebuild and now was the time to go. Gordon Clarke, who'd filmed the rebuild and accompanied the team to Rendcomb, was filming every part of the process on video, from reassembly to take-off. Now, in the passenger seat of an accompanying Stearman, courtesy of the Crunchie team, Gordon was shooting air-to-air footage of the tripe, shining and resplendent in its new scarlet livery. Gordon would later title his complete film of the rebuild and test flight: *A Living Legend*.

After twenty minutes above the Gloucestershire countryside, Robin brought the triplane in for a textbook landing – the first test having been completed with no problems to report.

'It's not the best day to fly,' he told Gordon grinning, 'but at least it flies!'

Somebody brought out a bottle of champagne and glasses to celebrate and the triplane was duly "launched" with a trickle of champagne poured over the prop. The test flight was a success. G-BEFR – or rather 425/17 – was back in business.

Robin sat on wheel of triplane following rebuild and just prior to first air test

It was possible to make proper plans now with the coming '95 season back on track. Robin talked of taking the triplane to Malta as he really wanted Sheila to experience his childhood home, to show her where he had lived and gone to school. By way of a recce, they took a short break on the island to check out the feasibility of taking the triplane. Robin was delighted to show her the various places he'd known, telling her the things he'd got up to and it was another great opportunity to cycle just as he used to do all those years before with Roger

Wilkin. This time a cold was making the trip a little less enjoyable, but nonetheless the trip was good fun, and yes, the possibility of bringing the triplane was definitely an idea to conjure with. On their return, he wrote a letter to the organiser of the Malta International Air Show.

It wasn't just the triplane that needed marketing that year; the Pitts also needed to earn its keep and so enquiries were made to various organisations including the organisers of the Tour de France as the tour was due to come to the south of England that year.

For the English leg of the tour he needed to write to Ashford Borough Council, the advertising department of Sport for Television Ltd, Kingston-upon Thames, to the media Marketing and Tourism department of the Portsmouth City Council, and to the advertising department of Halfords Ltd.

The rebuild had been a long and protracted process, costing far more in time and money than Robin had originally estimated, but it was to have its positive side, too. Robin, Dave and Ernie were in the process of establishing their credentials as renovators of classic aircraft and so when Vic Norman needed a Boeing Stearman refurbished aft of the firewall, he called Robin.

The work required that the Stearman be dismantled to the frame and then bead-blasted and powder-coated just as the triplane had been done. The fabric would be stripped and the internal structure inspected before being recovered, doped and painted; and all of this was to be done in the big shed of the Old Inn house. The whole process would take approximately four months over the winter season and help to keep the wolves from the door.

Illustration of Crunchie Stearman livery with handwritten notes (credit Aerosuperbatics Ltd and Robin Bowes)

The 1995 season looked even more optimistic, with Vic wanting Robin to fly the entire season solely for the Crunchies; he was no longer the reserve pilot.

The Crunchies also had a different colour scheme thanks to the re-spraying work performed by Robin, Dave and Ernie, and it was a livery in keeping with the silver foil wrapper of the Crunchie chocolate bar.

The triplane already had commitments, and for the first time ever, Robin would be free to fly other aeroplanes while confident that his friend, Dave Starkey would be able to fly the newly refurbished tripe in his absence. All in all, everything was looking very positive for the near future.

For 1995, it was intended that the triplane join forces with Nigel Hamlin-Wright's Avro 504K, which would be flying its first season on the Air Show circuit. For those organisers requiring a stunning WW1 display, the combination of the two would be a definite crowd-pleaser despite the logistics of them being based some 150 miles apart. The triplane was now based at Rendcomb. However, distance was no object and even the

Berlin historic Air Show at Johanasthal, was approached; only for Germany, the aeroplanes would be brought in by road rather than by air.

Robin was eager to promote the Avro at every opportunity; and in addition to the new kid on the block the triplane was already established in conjunction with the trio of three-quarter-scale replica SE5s' owned by Chris Mann, Doug Gregory and Des Biggs. Anthony Hutton's Fokker D VII added another famous German fighter and together the collection made for an excellent flying circus of WW1 veterans with the sole purpose of recreating the legendary dogfights of the Red Baron and the British aces.

In the early nineties when air show organisers were finding it difficult to afford battle re-enactments without sponsorship, the offer of a WW1 show from a collective had great appeal. Pilots Robin Bowes, Anthony Hutton and Tony Bianchi owned between them a substantial and varied collection including two Dr1 triplanes, a Sopwith Camel, Nieuport, Morane, Fokker D VII and Fokker Eindecker and a DH5. Add to that the SE5s' and the addition in 1995 of Nigel Hamlin-Wright's Avro 504K made for a very comprehensive flying circus!

With routines derived from earlier "spur-of-the-moment" shows, it made good economic sense to offer the collection as a package of carefully rehearsed aerial manoeuvres that could be tailored according to the organiser's budget. Each aeroplane would also be an attraction in its own right performing solo as well as in a choreographed dogfight slot.

If the WW1 aeroplanes were to compete on price with the RAF's team supported aircraft such as the Harrier that were available to larger organisers with full MOD support for very small money, then such amalgamations between pilot-owners were very necessary indeed.

On the 10th of February 1995, following a telephone

conversation with Mike Smith the organiser of the upcoming National Trust centenary celebration picnic at Stourhead, Wiltshire, Sheila sat down at her WP to write to him a letter of confirmation that the triplane would be available to fly on the evenings of 19th, 20th, 21st and 22nd of July for the sum of £1,400 plus VAT to cover all four evenings. She enclosed a brochure and a set of commentary notes to give more information on the aeroplane, and that she could probably come up with a music tape of Snoopy and the Red Baron, if required. She signed the letter, popped it in the post that evening and thought no more of it. It was just another booking for the diary that was filling the 1995 schedule.

And then another triplane appears

To help support the Army Air Corps museum at Middle Wallop, it was considered that an added attraction might be a collection of fixed wing aeroplanes. And so at Easter 1994, a triplane replica that had been built by Viv Bellamy – this time to Ron Sands plans – took its place in the museum in what was hoped to be the first of a few. The replica was based on the highest scoring triplane of the German Imperial Air Force No.45017 and authentically painted black. Only later when considering the budget and taking into account the value of such replicas did the Army have second thoughts on putting together a collection. This particular aeroplane would be the only fixed wing example that the Museum's committee would be able to purchase.

Robin had become aware of the Army's acquisition and he happily advised on what needed to be done if the aeroplane was to be flown as a display aircraft. Nick O'Brien was the only serving Army Air Corps pilot who was not so perturbed about flying such an ancient box crate and agreed to take on the challenge following refusals from his fellow flying officers.

Nick had very little information concerning the triplane; just one half of a sheet of A4, so he made his own individual approach to Robin for two reasons: firstly, he was concerned that he may be viewed as muscling in on Robin's pitch, for which Robin had worked hard and invested considerable sums over the years. For a salaried pilot to come in with a similar machine with the objective of making money for the Army Air Corps museum might not be viewed so kindly by some. Nick wondered if they might come to some accommodation in this respect? And secondly, Nick needed to know as much as Robin could impart about the madness of flying a triplane replica. What were its vices?

To Nick's relief, Robin was not at all concerned with any possibility that his pitch might be compromised by the arrival of another triplane on the circuit. He wrote to Nick putting his mind at ease and that to the contrary, another triplane would only add to the World War Ones already on the circuit.

The two pilots met at Rendcomb on the 25th of June 1995 at one of Vic's famous barbecues and flew both triplanes. Nigel Hamblin-Wright also attended with the stately Avro 504 while Chris Mann, Des Biggs and Doug Gregory brought in the SE5s for what turned out to be one of the best turnouts of WW1 replicas ever seen in the British Isles.

Robin was interested in sharing a forthcoming event with Nick – the National Trust Centenary Fete Champetre Celebrations at Stourhead Gardens. Robin would not be able to do the third night of the four night celebratory event and would be pleased if Nick would consider doing it with his triplane. He also suggested they could both fly on the Saturday evening as part of an exciting climax to the 20th century theme.

Chapter 10 – Inquest

… how everything turns away
Quite leisurely from the disaster; the ploughman may
Have heard the splash, the forsaken cry,
But for him it was not an important failure
(Musée des Beaux Arts), W.H. Auden

The morning of Friday June 14, 1996, was warm with the temperature rising, though cool enough in the narrow, white-walled Wiltshire Coroner's office. From the street, the office building was unremarkable in that it was not easily distinguishable from the many other pediment Georgian houses that form the ranks of shops and offices in Salisbury's city centre. Altogether, the historic brick buildings are a harmonious blend of, not only Georgian, but also medieval gabled houses and Tudor inns. Had Robin died just over the border in Dorset, then this would be Dorchester; or in Somerset, then it would have been Taunton. Indeed, so close was he to crashing at the point where all three counties meet, that the authorities would have been hard pressed to agree exactly into whose county he had fallen.

Salisbury lies within the central southern plain that saw the arrival of the first tribes to arrive in Britain from the continent just as the great ice age began to recede over ten thousand years ago. The modern cathedral city stands at the junction of four river valleys: the Avon and Bourne to the north and east, Wylie and Nadder to the west and south respectively. Sheltered by the rolling green plain which bears the city's name, Salisbury has been listed in countless tourist guides as "one of the most beautiful cathedral cities in Britain," and consequently enjoys a constant flow of tourists both native and foreign. On any day, but most especially during the summer months, every conceivable accent in the English language can be heard in the narrow streets and particularly around the 13th century cathedral grounds. Such accents are in addition to the many European, Arabic and Asian tongues that also fill the air.

Less than 80 miles from the nation's capital, yet so distinctly distant, Salisbury is regarded in the eyes of some, as being typically Wessex – the land of the West Saxons. Wessex as King Alfred the Great would have known it; and Wessex as Thomas Hardy would have known it. If such notable Englishmen were to return, we might imagine that once over the initial shock, there would be some recognition of the environment they had once known. Alfred would recognise his street planning, though not the traffic and the one-way system. Hardy might stand aghast at the sight of such expansive glass that forms department store windows and wonder why multi-coloured boxes on wheels that bear little resemblance to the few cars he knew, seem to take precedence over the pedestrian. Otherwise, both would know their location. They would certainly know it by the native's accent: on the cusp – mixed vowels that reflect the flat "a" and burrs of the West Country, though spoken in a manner which lends itself more to the pronunciation and sentence structure of the South-east.

Outside the Coroners Office in Castle Street, a small band of journalists from TV, radio and print, were already assembling for shots and sound bites of those witnesses arriving. Robin's death had meant something to the media – the man who flew as the Red Baron. What verdict would the coroner return? Would Robin's friends and family be satisfied with his verdict?

Inside, there was just enough room for the closest of Robin's friends such as Sheila and Ernie, his family – Christine, Harry and Violet, and also Wendy, Angela and Barry. Those witnesses that had been closest to the crash were also there: Paul Tolfree, Hilary Higson, Andrew Richards, and those who attended the crash such as Police Sergeant Maurice Lound; Ian Renee, a divisional fire officer with the Wiltshire Fire brigade; Philip Gilmartin, the senior investigator for air accidents from Farnborough and his colleague Charles Coghill.

Sitting Quietly

For those who were not at the inquest, Helen Tempest would probably have given the most compelling evidence for the deep mystery surrounding Robin's death that night of 20th July 1995. Whenever she had worked with him, he had always gone to great lengths to explain to her exactly how he was going to execute their coming display. More than any other pilot, he would plan a routine well in advance, calculating every move with near military precision. Whenever she flew with him, he was always so meticulous with planning. Other pilots prior to a show would dutifully go to the briefing, note the crowd line and display. Robin, in contrast, would walk her to the centre of the crowd line and say,

"Right, the wind's coming from this direction, there are trees over there, a gap there, so we're going to take off into wind, turn, come in and do this, then I'm going to climb to height and then we're going to come round, make use of the trees, make use of that curve …"

He would sit quietly and sort out his display not once, but at every single place they went, because, he would say, "each area has a different layout". He was the only person she had ever flown with who did that.

Stourhead was different. Robin had not walked the grounds. He would be aware that the crowd had a limited view of the aerial action. Wooded bluffs surround the lake on three sides, and to the south looking out across Turner's Paddock, visibility is soon lost again with more trees crowning the tops and slopes of the valley. For an aerial dogfight display, Stourhead was a tricky venue, something that Robin would not have failed to notice. Standing on the paths that surround the lake, or even picnicking on the grass slopes between the lake shore and the Bristol Cross obelisk, crowds would have strained to see the display above, at least by normal air show standards. Therefore, tight circles would be the order of the day. Police Sergeant Maurice Lound of the Wiltshire Constabulary on duty that night said he soon lost sight of the triplane behind the trees.

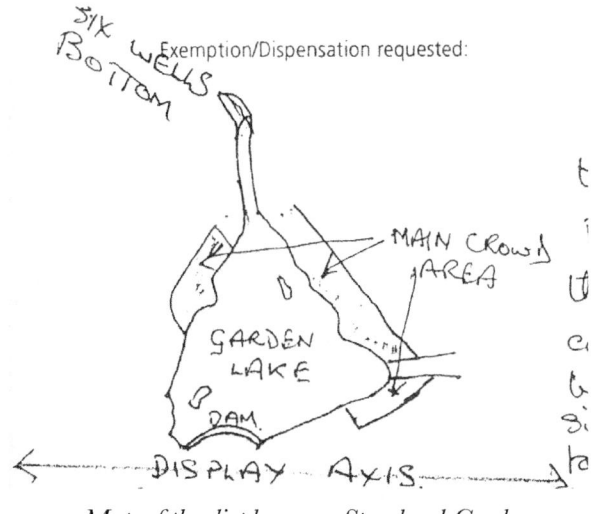

Map of the display area Stourhead Gardens

Robin's appearance as a solo act before the arrival of the SE5s', might suggest the arrival on stage of the wicked Baron in traditional panto style. The Baron, after all, was the anti-hero of the piece and so deserved the stage to himself for a few moments prior to the arrival of the good guys. Once the battle was done, after his smoky departure into oblivion, the SE5s' would have the skies to themselves before flying home to the "lines" – victorious.

In the gardens below, the fête champêtre spectators were part of the performance, too. Many dressed in costume to celebrate a particular era or decade sometimes in a manner that was significant to them. Perhaps the year Aunt Lucy won suffrage or the day Uncle Tom returned from the war with all but his leg and sound mind. Union flags were planted and the spirit of an earlier age pervaded as champagne and wine bottles were uncorked and wicker picnic hampers opened – their contents chosen especially for exposure: gleaming cutlery, dainty bone china upon which cucumber sandwiches and hard boiled eggs were placed; all carefully arranged on tartan blankets.

The Jiving Lindy Hoppers were dancing their way through the teenage years of the twentieth century; The Smallest Theatre in the World was performing "the history and decadence of a traditional aristocratic family" from its micro stage atop its transport – a motorcycle and sidecar. Altogether, more than 120 performers from singers, musicians, dancers, actors and cabaret artists, were getting into the swing of the celebration of the twentieth century.

According to the Witness Statement of John Christophers,[1] though it was never used at the inquest, the triplane appeared to be losing height as he and his parents first glimpsed it heading south across the lake – along the entry lane.

'There's the Red Baron now!' he exclaimed. This was the part of the fête champêtre he really wanted his parents to see; the reason he'd been so happy to return. His father, Arthur,[2] had been a navigator with a Mosquito squadron in the Far East during the Second World War, and *his* father –Arthur senior – had been an infantryman in the trenches of the First. Mrs Esther Christophers,[3] had been an ATS typist throughout

---

1   Name changed.
2   Name changed.
3   Name changed.

the war, and theirs has been a wartime meeting. Her father had also survived the bitter campaigns of WW1. To John Christophers, it was a certainty that neither of his 73-year-old parents had ever seen an actual triplane in flight before and certainly not in mock battle with three SE5s'.

The triplane was soon out of view – hidden by the trees along the lake shore – and the three continued their leisurely walk south toward the causeway that crossed the dam. Hopefully, thought John, they would reach the causeway at the far end of the lake by the time the Red Baron returned.

Also at the lakeside that night was the man who would be flying as the triplane pilot the following night. Nick O'Brien was at the northern end of the lake at the Bristol Cross obelisk stood with his son, Kevin, and fellow pilots Dave Griffin and Mike Smith – the air exhibition organiser who had corresponded with Sheila earlier that year. Griffin, a test pilot based at Boscombe Down, had himself only narrowly escaped death some weeks previously when the Spitfire he was flying crashed.

To John Christophers, Robin's entry into the arena was vastly different to that of the night before. For some reason, just as von Richthofen had exposed himself to Australian small arms fire by flying dangerously low across British lines to chase down a novice British pilot, so Robin also lost altitude as if he considered the lowest permissible altitude would give the greater advantage for the crowds trying to catch a fair glimpse of the triplane. And yet, if his level were to drop below the 500 feet minimum required by the CAA, he would contravene the display regulations and he would struggle to regain the height necessary for the dogfight. The triplane was no Tornado jet and would need to work hard if it was to make up the required altitude for the display.

In Turner's Paddock, Paul Tolfree, his wife Shin and their two children had settled down to their picnic on the steeply sloping Beech Knoll with family friends Major Andrew Ledger and his wife, Karry; their two children Thomas and George, together with Hilary Higson and her children Tom and Jayne. This was an excellent vantage point and as Turner's Paddock was owned by a friend, they weren't trespassing. They were particularly looking forward to seeing the air displays and eventually the fireworks that would close the evening.

Though much of the gardens were not visible from their position, they had an excellent view of the little scarlet triplane as it approached resplendent with its German markings, and were thrilled when it proceeded to fly right over their heads. However, there was nothing remarkable about the fly past – the triplane was on a level course according to Andrew Ledger – and there was no reason to suspect that there was anything wrong.

Nearby, Andrew Richards and a party of nine friends – all members of the Mid-Somerset Young Farmers Club – were also having a picnic. Andrew was particularly drawn to the flying display of vintage aircraft, because of his life long interest in aviation. The party had arrived at 7 pm to stake their claim in the Paddock and had been enjoying the operatic displays in the gardens before catching sight of the triplane. Like Paul Tolfree and Andrew Ledger, Mr Richards felt privileged to watch what initially seemed to be a private air display above them with the pilot dipping each wing in turn.

Walking along the gravel track that passes through the Paddock that evening, though not intent on visiting the fête champêtre, Mrs. Shirley Howe and her husband Ken, were walking as they often did, the first stretch of a route that circles the outer perimeter of the gardens when they heard a plane approach. Though, as with all their walks, they carried binoculars for bird watching, on this occasion they didn't use them. Shirley watched the triplane turn above them before disappearing from view in the direction of Alfred's Tower.

Mrs. Daena Hoare was at home with her husband, freelance forester, John "Jeremy" Hoare, just half a mile from the Gardens and the commotion of the fête champêtre. Daena was taking in the washing when she heard the sound of a light aircraft in the sky though she didn't bother to look up at it then. It had been a warm day and the washing was now perfectly dry. Jeremy, at the sound of the approaching engine, looked out of a downstairs window and watched the triplane pass over Beech Knoll. Everything seemed to be fine; a perfect summer's evening.

By the time John, Arthur and Esther Christophers reached the causeway that crossed the dam separating the main ornamental lake from the lower lake that takes the overflow, attendants were roping off the causeway in preparation for the air display. To John at that moment, the roping off of the causeway seemed a pointless precaution as, unlike an airfield, there was surely no crowd line as such? He wasn't to know that the roping off was simply for the entry lane and that once all four aircraft were in the display area, the ropes would be dropped and that they would enjoy a good view from this point. The night before, he had been on the eastern shore of the lake and remembered having a good view of the dogfight.

Based on his recollection from the previous night's viewpoint, he believed the display would take place above the spectators' heads – normally impossible at conventional air shows where pilots are not permitted under CAA rules to fly directly over the crowds. Whilst waiting for the Baron to reappear, he also noticed a handsomely built 5" guage model steam engine that had been running small children and oversized parents back and forth across the causeway. It was yet another example of a twentieth century scale replica of a once proud era in British history, and its presence was most apt on this nostalgic, celebratory evening. The driver, dressed traditionally for British footplate men of the steam era in blue denim overalls and black PVC topped peak cap, paid no attention to the people gathering at the rope, nor did he prepare to watch the display; instead, he just kept on tending to the idle, hissing engine, lovingly polishing, inspecting and seemingly besotted with the metal creation in the most worshipful of ways.

Mrs Hazel Davis and her daughter Elizabeth "Dibby" Rochford, together with her children Mark (10) and Kerry (9), had also reached the furthest point of their journey from the entrance gate and had passed the miniature railway just as the causeway was being roped off. They stood on an ornamental bridge to watch as the triplane approached at a height barely above the nearby trees. To Mrs Davis, the engine was making a loud noise.

John Christophers also heard the plane before he saw it. A tall man at 6ft 5in, he had a natural advantage as a spectator at any event and this was no exception as his parents and the few others that had gathered there strained to catch sight of the plane over a hedge. When he did eventually spot it, he was surprised to see the triplane so low – treetop height. The night before, it had kept a constant altitude of what he estimated to have been 800 feet.

To Shirley and Ken Howe, watching from the stone track that passed through Turner's Paddock, the triplane was now much lower than the first time they saw it only minutes before. In fact, it was so low that Shirley's immediate reaction was to "duck down" as it barely cleared the treetops.

On the top of the rise, the triplane now directly above them, Paul Tolfree, Andrew Ledger, Hilary Higson and their families watched incredulously as it first dived then levelled out before banking with its port wings raised skyward.

On the causeway, John Christophers saw it dive before recovering causing him to turn to his father for confirmation of what he'd just seen: 'That's a bit low for aerobatics, isn't it?'

Stood on the bridge, Hazel Davis saw the plane to her right.

Suddenly, it dived. To her ear, the engine seemed to have cut.

Nick O'Brien was watching the triplane turn when unexpectedly the nose pitched sharply. Turning to Kevin, he remarked, 'That's a bloody odd manoeuvre that?' To Nick's experienced eye, this was indeed a sharp "bunt". The triplane then seemed to recover before bunting again. From this steep dive, the triplane in Nick's words "described a sort of half circle as if flying around a cone."

In a nearby cottage, 19 year-old Suzi Steer was visiting friends. Sitting in the front garden, deep in conversation, she paid little attention to the triplane. What did cause her to take notice she believed – like Hazel Davis – was that the engine cut. In her statement to the police, this was a clear recollection for her and not shrouded in doubt. Suzi heard the engine literally stop – not splutter – stop. In her words: "The aircraft continued in a level flight for a few moments and then the aircraft's nose dropped, the aircraft began to lose height and as it did the tail of the aircraft rose."

Like other witnesses, she was under the impression this was part of the display, though her concern was that the aeroplane was too low. To Suzi, it even looked as if the pilot were intending to perform a "full roll" as she later described it, though she immediately dismissed the idea as "impossible" considering the lack of altitude. However, in using the term roll she stressed in her statement: "the pilot would have been on the outside of the roll throughout". She also described watching the triplane "invert" with the tail plane overtaking the nose and as it approached a large tree, continuing to roll and consequently "becoming more inverted". Like the witnesses on the knoll and Nick O'Brien at the lakeside, she also saw the wings turn "pulling the aircraft out of the roll" – Nick O'Brien's interpretation of "half circle round the cone". As the triplane banked sharply it caught the branch of a solitary oak tree on the lower bank of the Knoll like a fly catching its wing on the sticky edge of a spider's web. The resulting snap of fractured timber from both the tree and the wings sounded like a whiplash to those watching from the causeway and the ornamental bridge.

Depending on where the witnesses stood, sounds varied. Both Paul Tolfree and Andrew Ledger recalled silence after the "sound of tearing wood". From the causeway and the bridge, the noise of the triplane hitting the tree was like a huge whip cracking, and to Hazel Davis – like Suzi Steer – the engine seemed to start again. John Christophers on the causeway, together with his father, heard the engine scream as the plane climbed skyward. The pitch of the engine noise reminded him of the sound of a small radio controlled aeroplane screaming frantically to gain height. Peering over the hedge, Christophers saw the triplane climb and spin ferociously – he likened it to the type of high-speed climb and spin normally performed in aerobatic aeroplanes like the Pitts Special. Another witness said the plane seemed to climb and then "pirouette" before turning and falling nose first. Ironically, Robin would often have executed such an aerobatic manoeuvre in the Pitts. Climbing vertically at full throttle, spinning the aircraft in the climb before commencing a stall turn at altitude by kicking the rudder pedals to right or left from the stall thus causing the aeroplane to fall from its "pirouette" into a steep dive. It is possible that in a last desperate attempt to control the aircraft, Robin did just that.

On Beech Knoll, incredulity turned to horror as the families realised that the triplane was heading out of control and directly at them. The women looked to see where the children were. Then, miraculously, and with only microseconds to spare, the triplane veered away crashing nose first a mere 40 yards from them.

On such occasions, people witnessing the death of others react in extraordinary ways and do extraordinary things in what is at first an unbelievable happening. Friends from army days, Paul Tolfree and Andrew Ledger had both served in the Royal Hussars, but nothing in their collective army experience had prepared them for this catastrophe.

Immediately the triplane impacted, the two friends ran down the grassy slope and called to Robin to release his harness, but there was no coherent response. *Why had he been so low?* wondered Tolfree. Army training now instinctively in gear, they stopped in their tracks as the nose of the plane – its base embedded in the ground – ignited. Their desperate shouts at Robin drew no response. Fuel was already beginning to haemorrhage. It took just one little spark to set the plane alight. As the flames quickly took hold, there was no way the men could get to him. Tolfree even tried to grab a branch with which to beat out the flames, but behind, Hilary, Shin and Karry were in turn shouting at the men to get away for fear the plane would explode at any second. The children, who had been enjoying their picnic barely a minute before, were distraught. Seven year-old Lucy was hysterical.

Ken Howe ran from the foot of the hill to give assistance with Shirley shouting warnings not to get too close. She, too, feared it might explode. Like Tolfree and Ledger, he was beaten back by the heat of the fire.

On the causeway, John Christophers had lost sight of the triplane but heard the awful thud and saw, in what seemed an instant, a horrifying pall of black smoke and flames rising into the sky.

'Damn! Damn! Damn! Damn!' He repeatedly exclaimed; 'Oh, what a bugger! What a bugger! Poor bugger!' He heard someone nearby say, 'Oh, not the Red Baron!'

For some inexplicable reason, he reached into his inner jacket pocket, pulled out his press card and showed it to a young woman who was valiantly holding the rope her male colleague had just relinquished to run toward the crash scene.

'Let me through, I'm a journalist!' But he wasn't, and never had been, a reporter. Something, though, told him that he just had to be there.

Courageously, the young woman, who was probably no more than 19 years of age, and less than half his size, refused to be intimidated, steadfastly holding the line and, despite his towering frame, despite the possibility that he could so easily have brushed her aside, she rightly and courageously refused him permission to pass.

Turning to look again across the top of the hedge to the appalling black smoke and flame, John Christophers admitted to himself it was a lost cause. The noise of impact and now the resulting inferno convinced him that the pilot could not have survived. Had he run to the crash site, what would he have found? What could he have done?

From the bridge, Hazel Davis saw the triplane slide backwards before a row of flames began licking from under the wings. With the aeroplane well alight, Hazel Davis and Dibby hurried the children off the bridge. In Turner's Paddock, Hilary, Shin and Karry, gathered up their children and got them over the brow of the hill and away from the burning plane as quickly as they could.

For Nick O'Brien, too, the triplane had been lost from sight momentarily. The noise of the impact followed. Like Christophers, O'Brien couldn't believe what was happening and he looked for the most logical explanation: *Bloody hell, that was good! They've got ground pyrotechnics,* he thought. It wasn't unusual for display pilots to arrange ground pyrotechnics with event organisers. However, when the plane didn't reappear, he knew this was some serious pyrotechnic. The realisation of what had happened kicked-in when he heard someone exclaim: 'Christ! Yes he has crashed!'

The Baron is Kaput

The last major German offensive on the Western Front began in March 1918. In the see-saw balance of power that had risen and fallen on both sides in the aerial war and on the ground, the Germans were well equipped that spring with 3,668 aircraft battle ready on the lines. Despite this strength, they failed to capitalise and by April the offensive was all but over.

Von Richthofen being presented with the Iron Cross.

On April 21, von Richthofen recklessly chased a novice RFC pilot into the British lines at the Somme. He, too, was contravening the rules he normally lived by as he dropped to ground level in the pursuit and in so doing came under small arms fire from the bluffs of rising ground on which Australian infantrymen watched with incredulity. The first clear bursts at the triplane came from the Vickers machine gun of Sgt. C.D. (Cedric Basset) Popkin and Private R.F. Weston of the Australian Machine Gun Company. The triplane came just close enough for Popkin to fire off a seven-second burst which, taking distance and speed into account, he sensibly aimed ahead of the machine. Despite appearing to react to the burst, the Baron refused to give up the chase. Further up the Somme, Joe Hill and Ray McDiarmid, from their position on a high ridge, were witnesses to what was probably the Baron's fatal wound. To their left, Gunner Robert "Bob" Buie who, before the war was an Oyster fisherman from Brooklyn, New South Wales, initially could only stand and watch as the triplane passed directly overhead his position. Both he and sapper Tom Lovell – stood beside him – could see von Richthofen glance back but dare not open fire in case they caught the Camel. Further along the ridge Gunner William Evans opened up on the triplane followed by Buie once the aircraft was clear of Evans' position. The Baron had let off a burst of rounds at the Camel and strays were pummelling harmlessly into the ridge catching only pots and billy cans. With a clear view, Buie could see splinters fly and the triplane bank sharply. Evans repeated his fire and with a direct view into the cockpit he saw von Richthofen pull his goggles from his head and toss them overboard. Witnesses heard the engine scream at full chat before being cut – the sound replaced by the rush of air through the wings. No longer in full control, von Richthofen's scarlet bird wobbled and slew, quickly losing what little height it had. Sergeant Popkin saw it side-slipping before turning North-east followed by a spiral to the right, and then drifting to the left.

Like an angry wasp that had sustained a heavy blow from a rolled-up newspaper, it seemed to the men on the ridge to be turning back at them and opened fire one last time before dropping to the ground in a beet field beside the Bray-Corbie road. After bouncing and breaking its undercarriage, it slew round toward the road and came to rest nose down and tilting to its port wings.

Within minutes an Australian soldier had reached the plane under orders to take the pilot prisoner. He arrived in time to hear von Richthofen's last words: 'Das kaput'. By the time his comrades reached the plane von Richthofen was dead. Witnesses noted he had sustained facial wounds to the jaw undoubtedly caused by the impact with the Spandau gun handles awkwardly affixed just above the facia panel. There was also severe haemorrhaging from the torso, his fist holding the control column in a death-grip. Wounds to his legs and abdomen were clear to those peering into the cockpit – wounds that suggested shooting from the front, side and below the aircraft – not from above.

General Rawlinson's investigation of the incident concluded that von Richthofen had probably depressurised the fuel tank and cut his 100hp Oberursel engine in a last effort to prevent fire on impact; an act typical of a pilot of his experience and calibre. In so doing, the engine was stationery when the aircraft impacted with no resulting fire. Considering the evidence of von Richthofen's injuries, the General was also sure that Evans and Buie had most certainly fired the shots that mortally wounded the pilot, and he duly sent them a congratulatory message, though neither man ever received a commendation or war service pension from either the British or Australian governments. Nor did Buie – a quiet, modest man – make capital from it. Upon his return to New South Wales and fishing, he rarely mentioned his significant part in history and died aged 70 in 1964 with few but his family knowing his story.

Once von Richthofen's body was removed, enlisted men and officers helped themselves to souvenirs. Private Wormald, who together with an infantry officer lifted the body from the cockpit, found a card in the pilot's breast pocket identifying him as Hauptmann Manfred von Richthofen – the first official identification. Together with his mates, Wormald began to dismantle the wreck by cutting out of the top wing the black Maltese cross comprising some three square feet of red canvas. Such was the irony of this moment, for Richthofen, like many other pilots on both sides, viewed the collection of souvenirs as part of the victory and proof of successes in battle. How the tables had turned. Now he and his triplane were the most prized souvenirs on the Western front.

Death in the Evening

Horrified at what they had seen, John Christophers and his parents turned and walked away back along the same path they had just come. Hester noticed that the man with the steam engine continued to attend to the model throughout; not once did he look up, nor pay the slightest attention to what had just happened. He was engrossed. She was appalled at his lack of concern for the tragedy so near. After all, from what she had just witnessed, a man had most certainly died in a nearby field and in the most horrific of ways.

It wasn't just the steam engine driver who appeared to pay no heed. The party was continuing in full swing. People were laughing and chatting, sipping wine, nibbling nibbles, there was much meeting and greeting of old acquaintances. Blankets were spread out, picnics underway; flares stuck in the ground ready for lighting when the sun went down. Similar events were staged here each summer and a great time was had by all – always. Why should tonight be any different? Some seemed a little concerned that something had happened; but just what, no one seemed to know. Even the St. John's ambulance volunteers were stood at their station – seemingly bewildered. As John Christophers neared the exit gate, he was particularly irritated by the sound of laughing. Looking round, he saw a rather loud, large man laughing heartily for some inexplicable reason. It wasn't even a pleasant laugh. It was a laugh full of affect, a laugh of arrogance and disdain. Such a laugh could have attracted the devil, he thought. It repulsed Christophers, especially as the laughing man was in clear view of the rising pall of acrid black smoke. He wondered: *What was his excuse for being so hearty? A fellow human being had just that minute perished.*

Containing his emotions, Christophers led his parents out of the gardens. At 8.14 pm, just as they made their exit onto the lane that passes the entrance to the gardens, the first fire appliances and police cars were arriving, blue lights flashing, sirens blaring. A young man walking down the road, perhaps a member of staff seemed happy enough, smiling as he approached. His cheery demeanour received the brunt of John Christophers anger.

'What are you smiling about? Don't you know a man has just lost his life over there!'

He pointed to the rising pall. The young man lost his smile and walked past without stopping for explanation. It wasn't just him, or the fat man with the vulgar laugh; it appeared to be everyone. The party was continuing, bands, dancing, music, juggling, everything. DIDN'T THEY KNOW? To Christophers, it was not only the violent end that this poor man had suffered that was so ultimately tragic, it was the loss of his life in such public circumstances. Most people die in privacy; privacy afforded by their loved ones or health care professionals making them comfortable. Death is so often announced with solemnity, with comfort offered. That this was so public, and the public seemed either not to know or not to care horrified him as much as the impact itself. Was this the side of human nature that Brueghel knew when he painted the fall of Icarus? Was this, as Auden wrote: "… the expensive delicate ship that must have seen something amazing, a boy falling out of the sky, … and sailed calmly on."

In Turner's Paddock, Ken and Shirley Howe watched helplessly as the triplane burned. Circling above them, three bewildered SE5's. Robin's fellow pilots were now in clear view of the catastrophe below. Chris Mann flying G-INNY; Doug Gregory flying G-SEVA; and Des Biggs at the controls of G-BUWE.

Before leaving Stourhead, the Christophers family decided it best to let someone in authority know that they were witnesses. Stopping a member of staff, John asked where they could leave their names and was pointed in the direction of the estate office just yards from where they were stood and in a terrace row of picturesque cottages across the lane from the church and Spread Eagle Inn. Though it was not a time to admire it, the office (reached by a garden path), was in what was undoubtedly one of the quaintest and most typically English cottages he had ever been in. Through the office window he could see people still approaching the main entrance gleefully anticipating the evening's events. And among their number, staff dashing. What he could not see was the black smoke rising higher and higher at the far end of the gardens.

On leaving the office, he noticed a police car stopped some yards down the lane. A man was leaning in through the open passenger window talking to the officer inside. Christophers ran down to offer his version of events.

'I was a witness … he hit a tree, climbed and stalled …'

The man at the car contradicted him.

'He didn't stall.'

'And how do you know?' retorted Christophers with indignation.

'Because *I'm* the other triplane pilot.'

Christophers immediately apologised and withdrew.

'I'm sorry. So sorry!'

Shocked, Hazel Davis walked away from the ornamental bridge and, like the Christophers, wondered why other people seemed to think it was all part of the show. In the years to come, she would not return to Stourhead – not for this event or to walk the gardens, which she loved. Yet, to her, they were the most wonderful gardens of any she had ever seen here or abroad.

The Long Drive Home

Sheila was driving south along the A38 en route to her daughter Sarah's house. Sarah, recently married, had only been in the house for ten days and Sheila had yet to see it now that everything was unpacked, put in place, pictures hung on walls and essential items unboxed. Her attention, however, was consumed with the clock in the car's dashboard: twelve minutes past eight, twelve minutes past eight … Why did this moment mean so much to her, she wondered. It wasn't like her to be obsessed with time.

The Wiltshire Fire Brigade had received the first call at 8.03 pm and the nearest Fire Station at Mere was mobilised by 8.04. The first appliance from Mere, under the command of Sub Officer Pester, took ten minutes to arrive at the scene and was later joined by a second appliance from across the Dorset border in Gillingham. A third rescue tender on its way from Salisbury was turned back once it was realised that there were no survivors and that the body could be extricated without the need for additional rescue equipment. Both appliances were stood down by 00.54 the following morning.

Doctor Janet Millar was the duty doctor attending the fête champêtre that evening. Initially, stood near the main performance stage at the north end of the gardens, she was unaware of anything unusual until, in her words: "… a number of people drew my attention to a lot of smoke coming from behind some nearby trees."

Partly running, partly walking, she took the most direct route to the grassy hillside known as Beech Knoll. To her horror she saw the wreckage of the triplane well ablaze – its sole occupant well beyond her ability to give medical assistance. Indeed, so ferocious were the flames that she couldn't begin to approach. It was only when the firemen wearing breathing apparatus put out the fire on the triplane and the surrounding grass, that she was able to come close enough to certify death at 8.25 pm.

At approximately the time Dr. Millar was certifying death, Charles Coghill, a Senior Inspector of Air Accidents (Engineering) Air Accident Investigation Branch, Farnborough, was contacted at home by the AAIB Duty Co-ordinator. His instructions: to drive to Turner's Paddock at Stourhead and conduct an on-site investigation into the engineering aspects of this most unusual of aircraft accidents.

Fears confirmed

When Sheila arrived at Sarah's new home, she could neither relax nor settle. Sarah offered her a cup of tea, but for once, she refused; and Sheila was never one to miss a chance to chat with either of her daughters over tea. She made an excuse and left after only five minutes.

'It's no good Sarah, I've got to get home. I don't know why, but I've got to get home.'

When she arrived back at the Old Inn house, she routinely parked the car under the arch and put the kettle on as soon as she was inside the house; activated the answer phone and took a coffee jar from the cupboard. All routine stuff, but something wouldn't let her rest – some nagging feeling, though she couldn't quite figure out what it might be. Then she noticed someone approaching the house from across the Square – someone official – white shirt – quite tall. The loud knocking at the door startled her nonetheless – her heart thumping. The figure – an ominous presence – was at the door and he was a bringer of bad tidings. The happy days were over. This was the moment she dreaded and she anticipated its arrival. In the amount of time it takes to walk a domestic hall, to reach for a door handle and to open that door to a stranger who stands before you po-faced, not knowing himself just how to phrase the news that the man you loved with all your heart and soul is gone and will never come back.

Ever since she'd said goodbye that afternoon; the first flying lesson; the lack of any plan, the uncanny reason that for this one time, she didn't wave. At the opening of the door that dreadful summer's evening, it was as if in some other time and place God had shown her the script and told her: "This is where it ends, right here. These will be your last lines and they must be spoken and understood, for that is the part you play."

Waiting at Rendcomb

Seventy miles north of Stourhead Gardens, at Rendcomb airfield, Helen Tempest was beginning to wonder why Robin was overdue. It had already been a horrendous evening with her good friend Matthew Boddington crashing into wires with the Tiger Moth. Miraculously, he'd walked away from the burning wreck and she'd taken him to hospital for treatment to cuts and bruises. He'd be alright, but that was such a close, awful thing to witness – the aeroplane crashing and burning; it could have been so very different.

It was to have been a simple flypast by two Tiger Moths before leaving Rendcomb that evening. Now, it had all gone horribly wrong. She'd talked to his mother – her Godmother – on the phone to reassure her that Matthew was okay and that he'd be released from hospital soon. There was nothing more she could do so Helen decided it would be better to return to Rendcomb rather than go straight home. She couldn't rest and needed to come to terms with what she'd seen; she needed to help Vic and Smithy clear up the field. By 9pm, nothing remained of where the Tiger had crashed save for a grim circle of ashes.

As a glorious sun set in the warm, western sky, she looked up from the circle of burnt grass and spoke of her concern.

'Robin should be home soon. I wonder where he is?'

Such questions have no answers and the three speculated that for some reason he must be staying elsewhere, though it would be typical of him to ring soon and let them know. After all, it was not a day to forget about overdue aircraft.

Vital Evidence

At Stourhead, an investigation into the cause of the fire had begun as soon as it had been extinguished, although this was initially hampered by the necessity not to disturb vital evidence contained within the aircraft and scattered around the immediate area of the crash site – which was imperative if the air accident investigators of the CAA were to do their job effectively. With this in mind, Sub Officer Pester gave the order to tape off the crash site to prevent unnecessary disturbance. Indeed, there was a trail of small fragments of wreckage and broken branches from the tree – the top of which was some 70 feet above the crash site. Some fragments were from the lower left wing. Also, it was not possible to remove the body until the CAA had arrived to perform their investigation.

On cursory examination, and considering the evidence of witnesses such as Andrew Ledger, Paul Tolfree and Ken Howe, who had all seen the plane ignite after impact, it was considered that the most probable cause of the fire was a ruptured fuel tank or fuel lines with ignition coming from an electrical component such as the engine's magneto which, without being turned off, might have continued to spark after the engine had cut out. Lesser possibilities included a spark caused by friction or leaking fuel over a hot component. Another possibility, though much less probable, was that fire might have begun while the aircraft was still in the air, though no eyewitnesses reported seeing evidence of fire especially when the aircraft impacted. Without corroborating evidence, this theory was soon discounted.

As would be expected at such an event, first aid fire fighting equipment was well provided for in the main event area of the gardens and as recommended by the Fire Brigade, but Turner's Paddock was too remote to warrant extra equipment. And, even if equipment had been available, the time between impact and ignition was too short for a fire extinguisher to have been fetched and put into effective use.

Waiting for News

Unlike Sheila, Helen felt unable to return home and so as soon as the clearing up was done, she drove to Tanya Gaze's home to tell her of the awful events.

'Tanya it's been absolutely horrible, Matthew crashed into power lines, he's in hospital. It's just been the worse day.'

The two wing walkers talked at length, but had Tanya not decided to go to bed early, Helen might not have paid any attention to the television and News at Ten. She watched to see whether Matthew's crash had made national news, but it was a report concerning the death of another air show pilot at an event in Wiltshire that made her sit up. Although not mentioned by name, she knew it had to be Robin. Within twenty minutes there was a knock at the door; it was Smithy, uncharacteristically wearing a suit.

She knew Smithy so well; knew he never wore full suits. He quickly confirmed her fears – it *was* Robin, but she already knew it. Vic Norman had also heard the news from a television report. He phoned other members of the Crunchie team including Mike Dentith, but was unable to get a response on Helen's number. When he eventually arrived at Tanya's, the three sat solemnly in a row in a house that none of them had ever been in before, and agreed that they shouldn't wake Tanya because it was all so horrible. There was no point in another soul just sitting there in a state in the hallway.

Ironically, Robin had been contracted to spend most of the 1995 season flying with the Crunchies, whilst Dave Starkey deputised as the Red Baron. The contract meant five-day a week employment for that season, for at least four months from the beginning of May. For the first time since those early days as a car salesman in Plymouth, he would have been earning a regular weekly income.

The Initial Investigation

Charles Coghill arrived at the crash site at 10.45 pm and immediately carried out an initial investigation after which the body could be removed.

Softly As I Leave You

Over the next 24 hours, news of Robin's death was greeted with shock and disbelief by family, friends and acquaintances. That night, Sheila rang Ernie, Dave and Sarah. Gathered at the Old Inn house the friends talked into the early hours trying to find out what they could. After Sarah had gone to bed and Ernie and Dave had left, Sheila prowled the house in a daze. As 5 o'clock approached, she longed for a proper answer.

Turning on the radio for the local news, she heard the first bars of Matt Monroe's hit song *Softly As I Leave You*. It was if the whole world was asleep and someone somewhere was playing this sad, sweet melody just for her. When the song finished, she returned to the seat on the patio where she had been for most of the night and watched the most beautiful sunrise she had ever seen. It was pure peach with the grey of the trees on the skyline and a most beautiful new sun rising. This was the confirmation she needed: *You're okay*.

As she watched the sunrise, a man whom she had never met

and never heard of, was preparing to leave his home in Hampshire to travel the 80 odd miles to Turner's Paddock.

Word Spreads

The world of light aviation is a small one, and throughout that night and the following day friends and fellow aviators learned of the death of a dear friend. Tanya Gaze awoke to find a note from Helen: "Our little Robin has gone".

Nigel Hamlin-Wright couldn't believe the news and hoped in vain that someone had got the story wrong. He phoned his friend Mrs K who was dumbstruck. Both he and the indomitable Mrs K had fallen for Robin's charm and solid good nature. She was very aware of the part that Robin had played in the success of the Avro and the two had become almost synonymous in her mind. Suddenly, without his guiding hand, both friends felt very much on their own.

Across the Irish Sea, Liam Byrne rang Madeleine O'Rourke who immediately went to a local newsagent's to buy a British paper for confirmation. On the continent Phil Tongue flying as first officer on a British Airways Airbus, heard from a colleague that a triplane had crashed in England. As soon as he landed, he rang Jeff Salter in Belfast to check whether the triplane was Robin's. As the tripe was rumoured to be black, Jeff reassured his friend that it couldn't be Robin. Shortly after his conversation with Phil, the phone rang again. It was Jerry Snodden who had heard rumours that there had been a triplane accident and could it be Robin? In order to clarify once and for all just whose triplane it had been, Jeff called Helen Tempest at Rendcomb.

'Look, Helen, I know this is a bit of a silly thing, but there's a story going around here that there's been a triplane crash in England. Is it Robin?'

Helen confirmed his worst fears. Replacing the receiver, he

walked out of the control tower, climbed the grassy slopes of a nearby hill, and wept.

The Morning After

Mr. Philip Gilmartin, Senior Inspector of Air Accidents at the Air Accidents Investigation Branch, Farnborough, arrived at the crash site in Turner's Paddock on the morning of July 21st. His purpose: to investigate the operational aspects of the accident; to piece together the story of Robin Austen Bowes, aviator, and specifically his last flight in G-BEFR. There had been a police guard on the skeletal remains of the triplane throughout the night.

Whereas in 1917, von Richthofen's aeroplane had been quickly dismembered by souvenir hunters, it was imperative that for a modern forensic investigation to take place, every minute piece of evidence must remain undisturbed. It had taken two hose reels and just five minutes to put out the fire that had consumed the combustible elements of the triplane and the surrounding dry grass. With the scarlet canvas completely destroyed in the fire, Gilmartin's job in examining the airframe and flying controls was somewhat easier than it might otherwise have been. Friday was another fine weather day with clear blue skies and rising temperatures. Too warm to be walking around in a white one-piece coverall suit – a necessary uniform item of clothing that prevented stray fibres from the everyday clothes of the investigator contaminating the evidence.

Salvaging History

After von Richthofen's crash, the interest in the downed triplane soon spread with several parties having a vested interest in claiming what was left of it. The Red Baron – as von Richthofen was now known – had, through the war years, become a celebrity, though an anti-hero in Allied eyes. During 1917, the British government were preparing to establish a National War Museum and, in need of exhibits, custodians-to-be were requesting commanding officers to collect and preserve items of historical interest.

Orders to salvage the Baron's triplane fell on the lone shoulders of Lt. N. J. Warneford, Equipment Officer with No. 3 Squadron Australian Flying Corps. He located the wreck at 2 pm on April 21, just three hours after its crash landing. Prudently deciding to wait until nightfall and a lull in shelling, the remains of Fokker Dr.1 no. 425/17 were removed to an airfield at Poulainville. Had it not been for its most wanted pilot, the wreckage may well have remained where it was until the end of the war. After all, there was nothing about it that warranted losing life to a shell barrage under normal circumstances – any other triplane would have been abandoned to its fate.

Warneford's report described the aircraft – quite understandably – as "a complete wreck" and "knocked about by shell splinters" – this in addition to the damage caused by the small arms fire and subsequent crash. According to Warneford:

"It is painted bright red all over. Date on top plane 13-12-17. Fabric of rather better quality than usual. Finish of engine is better than those captured in previous machines of this type."

A considerable amount of fabric still covered the triplane despite the cuts and tears inflicted by the souvenir hunters. The propeller was splintered and most of the instruments had been pulled from the facia; the gunlocks and ammunition were also missing.

All in all, there was little left to preserve, but components were labelled and eventually dispatched to those who put in a substantial claim on them. Captain Roy Brown requested the engine, but initially received nothing. Museums and air force squadrons in Canada, Britain and Australia acquired most of the remaining salvageable parts though as time went on and a second world war took precedence in people's lives, items were lost.

One of von Richthofen's triplanes did survive the First War; however, only to become a casualty of the Second. Preserved by post war German governments and highly prized by the Nazis and the Luftwaffe, a Fokker D.III biplane and a Dr.I triplane, Nr.152/17, were displayed in Berlin's Zaughaus until ironically, their destruction by the RAF in a bombing raid. In the years between and after both wars, pieces of no. 425/17 have become as sought after as relics from the true Cross, and seemingly as plentiful. Indeed, so many people claimed to have the original manufacturer's nameplate at one time, that it was jokingly suggested that the triplane had not been shot down, but fell under the weight of the brass it was carrying!

Leaving a Tribute

By mid-morning, television news crews from Bristol and Plymouth had arrived and were granted access to film the wreckage whilst Gilmartin and Coghill painstakingly went about their work. One cameraman shot a gruesome and evocative shot of a pair of goggles and mask lying isolated in the grass and seemingly untouched by the fire. When John Christophers saw it on the regional news later that day, he was shocked and appalled and immediately phoned the news desk at the station where upon he railed at the producer. He'd seen the plane burn. How could anything have survived? What would the pilot's family feel if they were to see this? Such personal items! What had journalism come to if reports were to stoop to this gross sensationalism?

Christophers had also returned that morning, though not to visit the crash site. He left flowers with National Trust staff requesting that they be placed in the Church in memory of the "brave pilot who died". He wasn't to know that this act of condolence was to be the only spontaneous tribute from a general public of some 5,000 who had been in attendance the night before.

He was also aware that the crash had formed something of a metaphorical circle in respect of his father's life. In 1944, Arthur Christophers, a navigator with a Mosquito squadron operating in the Far East, had watched helpless as his Canadian pilot and an engineering officer were killed when their Mosquito crashed shortly after take-off. Arthur had been ordered to stand down for what should have been a short, routine flight that would allow the engineering officer to ascertain whether maintenance work to one of the twin Rolls Royce Merlin engines had been satisfactory. The crash was the third in what had been a spate of similar incidents where the "wooden wonders" – recently introduced to tropical operations – were experiencing what appeared to be structural problems. No similar incidents were occurring to the type operating in cooler climates.

This particular accident was enough to cause the Mosquito's maker, Geoffrey de Havilland, together with C.J. Chabot– an expert of renown in the world of aviation – to fly out to India at the first opportunity and investigate the crash first hand. According to Chabot, speaking in a BBC television documentary about his life and broadcast in the late 1970s, this was the first air crash investigation of its kind. The result of that investigation concluded that the formaldehyde glue used to adhere the wooden airframe was not sufficient for tropical conditions due to its tendency to molten.

Post Mortem

At Salisbury District Hospital, a Post Mortem examination was underway by 11.00 a.m. and conducted by Dr. Henry

Drysdale – a civilian consultant in Forensic Pathology from the RAF Institute of Pathology and Tropical Medicine. Police Sergeant Lound identified the body as the one he had seen in the cockpit the evening before, though the official identification was reliant on dental records.

Removal

Later that day, the remains of the triplane were removed to AAIB headquarters at Farnborough where, over the following six months, it would be examined in greater detail whilst witness accounts and flight records were collated.

The Reports

By February 22, 1996, the AAIB investigation was complete and the findings were published in the AAIB Bulletin in accordance with protocol. The investigation reports into the incident are reproduced here in part by kind permission of the AAIB and Messrs Coghill and Gilmartin, the report's authors.

> "As the aircraft commenced the run in for the first manoeuvre, it was seen to have a disturbance in pitch, followed a short time later by a very sharp and uncharacteristic pitch nose down to an almost vertical attitude. During the ensuing dive, the aircraft was observed to either wing rock or complete one or possibly two rotations. The aircraft was banking to the left as it recovered from the dive and disappeared from the view of most spectators as it went behind some trees away from the gathered crowd. A small number of video recordings were obtained from spectators and examined by AAIB. Only one short sequence showed the aircraft in flight just before the impact. The initial pitch down manoeuvre was not

seen. The aircraft was seen pulling out from a dive but steeply banked to the left. There was an indication in some video frames that the right aileron was up (roll right) but the left bank angle was maintained until the aircraft disappeared amongst the trees. The images of the aircraft were small and indistinct but they gave no indication of any major structural failure.

"The aircraft collided with the top of a tree and spun to the ground, where a large fire engulfed it almost immediately. Two witnesses close to the impact point attempted to rescue the pilot but were prevented from doing so by the intensity of the fire.

"A post-mortem report showed that the pilot died as a result of the fire. There was no evidence to suggest that pre-existing disease, alcohol or drugs had any part in the accident…"

The engineering examination conducted by Mr. Coghill, stated:

"It [the triplane] descended on to the ground upright but banked to the left and in a nose down attitude with no horizontal speed. … In descending 70 feet after colliding with the tree it had covered a horizontal distance of 160 feet. Although its final descent had been vertical it had substantial forward speed on hitting the tree when its descent angle had been quite shallow. The aircraft had been rotating when it hit the ground and it came to rest facing back towards the tree with which it had collided…

"The bottom of the fuel tank was punctured in the ground impact… The welded steel tube structure of the fuselage, tail and ailerons survived the fire, as did the flying controls and main structural fittings. All the main structural attachments survived and in these and in the structure that remained no pre-crash failure was found apart from the damage to the bottom left wing caused by impact with the tree.

"There were two failures in the flying controls. A control rod in the elevator system, which passed underneath the pilot's seat, had broken. The rod end had failed in bending overload when the pilot's seat had collapsed downwards onto it during the ground impact.

"The second failure was in the top rudder hinge. The rudder was supported on two hinges each of which was attached to the fuselage by a steel strap. Each steel strap was wrapped around a hinge bearing on the rudder vertical spar and welded to its outer race. The ends of the straps were attached to the stern post of the fuselage frame by through bolts. Both sides of the strap on the top hinge had fractured. Each fracture path passed through the bolt hole, aft of its centre. The strap and the surfaces of the two fractures were covered by soot and burned paint or fabric. The hinge had failed in a fashion consistent with the ground impact but metallurgical examination showed that both sides of the strap had suffered fatigue cracking.

"The fatigue had developed as a result of reverse bending loads, the result of repeated side loads on the rudder in flight. The extent of pre-existing cracking through both sides of the strap was such that the strap was virtually severed by the fatigue. This makes it likely that the hinge had become detached in the air before the aircraft hit the ground. The right side strap had also cracked further aft, where it was welded to the bearing's outer race, but it had not separated at that location.

"The metallurgical examination revealed that there

were two material defects which would have lowered the strap's resistance to the development of fatigue cracks. There was a reduction of the carbon content of the steel in the strap's surface layers and there was some microscopic oxidisation pitting of the surface (from which cracks had initiated). Both effects could have occurred if the strap had been heated when it was being formed into shape to fit around the hinge. However, the decarburization [sic] of the surface could also have been the result of the manufacturing process for the steel sheet from which the strap was made. There was also some corrosion pitting which was the result of slight deterioration with time.

"Such a failure of the rudder hinge did not appear to explain fully the eye-witness reports which described the aircraft pitching downwards at the start of the accident sequence with no sign of any yaw. It did not appear likely, because of the amount of clearance available, that the rudder would have fouled the elevator when the top hinge broke. There was no other structural or control failure which could have caused such a pitching manoeuvre.

"As a result of some witness statements the possibility of an engine power loss was considered. Both (wooden) propeller blades had broken in the impact, primarily in rearward bending, with no rotational damage. However, the fact that both blades were broken showed that the propeller had been rotating at the time of the first tree impact. Even so, the rotational evidence was not strong enough to indicate that the engine was producing power. It is possible that the pilot may have closed the throttle as part of his attempted recovery from the aircraft's steep dive or carried out a crash drill before impact.

"The engine was removed and stripped but no

mechanical failure was found in it or its accessories. The severe heat and impact damage could have destroyed or obscured evidence and so it remains a possibility there may have been a power loss in flight. A fuel sample from the supply from which G-BEFR had last been refuelled showed a deviation from specification in existent gum, which may have come from plasticisers [sic] in the fuel hose but this was not thought to be significant.

"… During the final Annual Inspection (about 20 operating hours before the crash) the rudder hinges had been subjected to a visual inspection for security and wear but had not been dismantled. The rudder hinges fitted to the aircraft at the time of the accident were different to those shown in the Redfearn plans. The original hinges had developed cracks and the hinge layout was changed while the aircraft was being re-engined and rebuilt over a two-year period following the crash in 1992. The aircraft had completed 54 operating hours since the rebuild. Neither the rebuild report submitted to the PFA nor the other aircraft documents contained any record of the rudder hinge modification or of it having been approved by the PFA."

At the inquest, Mr Gilmartin produced his report and that of a flight test report compiled by test pilot R. D. Cole flying a similar replica Fokker triplane. Cole's report on the triplane replica aircraft type was complimentary overall and part of which is reproduced here by kind permission of the report's author. The report comprises the results of six tests: Cockpit Assessment; taxi, take-off and landing; climb; stalling; handling; acrobatics. The following paragraphs come under the heading of Conclusions and Recommendations, which complete the report.

"A quite delightful experience for any pilot, mainly because of the aircraft's relatively high power allied with relaxed stability. Overall, however, this is not a machine for the uninitiated because of the lack of view on the ground and the positive approach required to taxi, particularly directionally.

"Airborne the aircraft is easy to fly as long as the pilot remembers that the rudder is almost the primary control!

"The handling qualities of the aircraft are acceptable for the recommendation of the award of a Permit to Fly including limited acrobatics."

R D Cole
Test Pilot

Questioned by the Coroner, Gilmartin spoke of Robin's considerable experience particularly in respect of the triplane. He also explained in detail some of the more technical aspects of the report. The following is an abridged summary of comments taken from this author's notes and Mr. Gilmartin's written statement to the Inquest. It is not quoted verbatim.

'Robin Bowes learned to fly in 1967 gaining his Private Pilots Licence after training at Roborough Airport, Plymouth. He was issued with a Basic Commercial Pilots Licence in 1990. This was restricted to display flying only. He first flew the Fokker triplane G-BEFR in April 1984 and purchased it for the purpose of display flying which he commenced later that month.

'The aircraft suffered an engine failure in June 1992 whilst flying in Germany and had to make a forced landing, during which it sustained major damage. Once recovered to Mr Bowes home in Devon, the aircraft was rebuilt and the engine replaced. It next flew in March 1994. Since that time, Mr Bowes had flown over 48 hours in the aircraft, the majority of this time in display activity.

'In respect of the fête champêtre at Stourhead, the planned display sequence involved G-BEFR arriving at the display site in order to commence a series of gentle aerobatic manoeuvres. G-BEFR would then be joined after a few minutes by three replica SE5a biplanes in the display area and stage a mock dog fight between the four aircraft. The culmination of which would see the SE5a aircraft as victors. The display organisers received the necessary approvals from the CAA for this particular event with a nominated display line greater than the minimum 150 metres from the main spectator area in respect of aerobatic aircraft. However, some spectators such as those in Turner's Paddock were not within this area.

'On the morning of July 20, 1995, Mr Bowes returned to Dunkeswell to work on the engine exhaust of G-BEFR. It was a particularly warm day with several hours of uninterrupted sunshine recorded. Mr Bowes was seen to work for several hours and showed some external effects of exposure to the sun. The aircraft was refuelled with 49 litres of Avgas from the airfield's refuelling facility. Mr Bowes also took an evening meal at the airfield around 1800 hours. He took off at 19.25 hours with a night stop kit stowed and also a tool kit. It was his intention to land at Rendcomb, not Dunkeswell, at the end of the display.

'G-BEFR appeared in the vicinity of the site about two minutes early and made a wide gentle circuit around the area awaiting the 20:00 hours start time. He was in RT contact with the other aircraft on a pre-arranged frequency and was heard to say, 'Running in, one minute.' Which was his last and only call. Witnesses noticed that as the aircraft commenced its run in for the first manoeuvre, it had a high angle of pitch ...'

Gilmartin was asked by the Coroner to explain what is meant by "pitch".

'Pitch attitude relates to the nose of the aircraft.' Gilmartin resumed his statement.

'This was followed by a very sharp and uncharacteristic pitch which put the aircraft into a dive. The aircraft was observed to either wing rock or complete possibly two rotations. Recovering from the dive, the aircraft was seen to bank away to the left and disappeared from view of most spectators going behind some trees. A video recording shot by a spectator shows the right aileron up for a roll right but the left bank angle was maintained. There is one aileron on each of the top wings to control the aircraft in a roll. For the right aileron to be up, it would indicate that the pilot was trying to push the aircraft to the right, but it actually kept going to the left. The aircraft was then lost to view behind some trees. Neither the video recording nor eyewitness accounts give any indication of structural problems. It should be noted that the images of the aircraft in the video were small and indistinct. The weather was fine with a light westerly air stream and the temperature was in excess of 25 degrees Celsius and therefore shouldn't have caused him any problems. The evidence would point to him not being in control during the manoeuvre.'

Asked about the rudder, Gilmartin explained:

'The rudder is very powerful on this machine and can prompt the aircraft to roll. The triplane is unusual in that it has an all-moving rudder as opposed to a vertical fin. Its primary effect is to control the aircraft directionally. Had it failed, it would have given the pilot a significant problem. If there was a major control problem he would have attempted to find a suitable place to land.'

And in answer to why the aircraft should make a sudden dive:

'You increase stability by increasing flying speed. The pilot may have considered that there was an urgent need to increase speed, and therefore pushed the stick forward to instigate a dive. He then tried to pull out believing he had enough time to do so. I believe that he was aware that something was wrong and did not have the time to analyse what it was.'

Following on from Mr. Gilmartin, Charles Coghill addressed the inquest with his findings of which for the most part mirrored those of his esteemed colleague: that it had been the failure of a steel hinge strap connecting the rudder to its hinge post. As with Mr. Gilmartin's statement, Coghill concurred that:

'Initially I considered the break a result of the accident, but then a metallurgist found fatigue cracking in both sides of a metal strap, which is wrapped around the hinge. Such fatigue cracking is progressive and can be caused through repeated loads. These cyclic, or repeated loads cause the fatigue crack to grow thus increasing the local stress ending in the rapid destruction of the strap. The hinge probably came apart in the air. The metallurgist reported that there were two material defects, which could have lowered the strap's resistance to the development of fatigue. The first being a reduction of the carbon content of the steel in the strap's surface layers and secondly there was some microscopic oxidisation pitting of the surface. The Metallurgist considered that the decarburization [sic] of the surface could have been the result of the manufacturing process of the sheet steel from which it was made, or it could have resulted from the heating process when the strap was being shaped to fit around the hinge.'

Asked to explain the purpose and number of hinges, Mr Coghill replied: 'The rudder is supported on two hinges attached to the fuselage by a steel strap held by a bolt. It was the top hinge that had broken due to fatigue cracking.'

Asked why the cracks in the hinge had not been identified when the aeroplane was inspected following its rebuild, Coghill replied:

'Fatigue cracks, especially when they are painted over, can be difficult to detect on examination. The plane doesn't need to be dismantled. It is left to the inspector.'

Coroner: 'Could you tell the Inquest, Mr. Coghill, in what condition was the bottom hinge after the accident?'

Coghill: 'The bottom hinge was bent to the left but with the rudder still retained.'

Coroner: 'Would the failure of the rudder hinge explain the aeroplane's sudden dive?'

Coghill: 'It is difficult to understand why he entered into a steep dive with no sign of any yawing. He had managed to affect a recovery but there is the effect of the rudder. It is difficult to connect the failure of the hinge with this manoeuvre. It is possible that he might have thought something else was wrong which a dive would have put right.'

Coroner: 'Would you explain what is meant by the term "yawing"?'

Coghill: 'Yawing is when the nose of the aeroplane turns to the right or to the left in a horizontal plane.'

Coroner: 'Some witnesses have stated that they did not hear the sound of an engine immediately prior to the crash. Would engine failure explain the sudden dive, or pitching downwards, especially where there was no sign of any side slip movement which surely rudder failure would have caused?'

Coghill: 'I considered the possibility that there may have been an engine failure which would have resulted in a loss of speed resulting in a stall. There was evidence that the propeller was rotating when the aeroplane impacted with the ground, but there was nothing to indicate whether the engine was producing power on impact. It is possible that the engine stopped when the aeroplane hit the tree or that the pilot may have closed the throttle as part of his attempted recovery from the dive. I removed the engine and dismantled it, but found no mechanical failure in either the engine or its accessories. However, evidence could have been obscured or destroyed as a result of severe heat damage resulting from the fire. I have analysed the daily fuel sample at Dunkeswell aerodrome where he refuelled, and found that it failed on two counts – evaporation together with a high level of existent gum due to deterioration in the hose between tank and nozzle. It could cause an engine problem in the carburettor, but it's not relevant for this accident. I didn't think it was significant. However, it remains a possibility that there may have been power loss in flight.'

Coroner: 'What effect would the loss of the rudder have on this aircraft in flight?'

Coghill: 'The rudder affects the directional stability of the aeroplane. On the triplane, the large rudder would have moved in an unusual way due to the securing of the lower bolt and cables. This doesn't explain the angle of pitch in the initial steep dive, and I can't connect the two. A possible explanation is that the directional stability of the aircraft is dependent on the keel area behind the pilot. The rudder on the triplane is in such a configuration that without a fixed tail fin, unlike a modern aeroplane, the loss of rudder control would cause instability. Therefore, the pilot would need extra speed from a dive to aid this stability and affect a forced landing. With the loss of rudder control, I would expect even a very experienced pilot to have severe difficulties.'

Coroner: 'Assuming the pilot had a defective rudder, would he have had enough control just prior to the accident to steer the aeroplane away from people situated in Turner's Paddock?'

Coghill: 'There is a real chance that he was steering in an effort to land the plane without risk or harm to others and in particular to those he could see on the ground.'

Asked by the jury whether the pilot would have had any rudder control after the breakage of a hinge, Mr. Coghill replied:

'Operating the rudder pedals could have produced a waggle as the cable was still connected.'

And questioned by Ernie Hoblyn about the condition of the remaining hinge, he replied:

'The bottom hinge was bent. It had been bent as a result of either the impact, or by further stress placed on it by the top hinge. It was a single bend consistent with impact damage. There was no reverse bend and may have been operating to some extent. The fracture most probably occurred in flight.'

At the coroner's request, Mr. Coghill assured the Inquest that in light of his investigation he would write a letter to the Popular Flying Association making reference to the cracking in this type of hinge, especially as the same layout was to be found on other aircraft. And, that as no record of a modification had been made, he would remind the PFA that it should point out to its members the importance of their notifying the Association of all modifications.

Sheila, like Ernie, was not so sure that Robin had died as a result of a hinge failure. There was something else on her mind and ever since that awful night of July 20, 1995, she had remembered an incident that may have had a greater bearing on his eventual downfall. Asking permission of the Coroner to make a statement to the Inquest, she recalled a visit to friends nine months before he died.

'In October 1994, Robin and I had gone to our next door neighbours for a meal and where we drank a couple of bottles of wine, although Robin was not a drinker by any means. Our neighbours wanted to show us a conversion they'd made upstairs, and so we were taken up and whilst standing on the landing looking down, Robin passed out. Fell flat on his face. He was out cold but within a couple of seconds he was up, although rather woozy. It was a momentary loss of consciousness. He sat there for a few minutes wondering what had happened, quite unaware. We all wondered whether it was too many glasses of wine, and as I said, not being a drinker, it probably affected him.'

The Coroner asked her whether Robin had reported the incident to a doctor.

'He didn't discuss it with the doctor as far as I'm aware. I kept an eye on him, but nothing came of it.'

'Had he suffered a similar problem prior to his collapse when at your neighbours?'

'Nine days previously – also a hot day – he'd gone into a dive.'

'Mr. Gilmartin has said at this Inquiry that he discussed Mr. Bowes health with Dr. Drysdale, who performed the Post Mortem examination, and with Mr. Bowes G.P. He also discussed this with you and the pilots at Rendcomb, but there were no reports of Mr. Bowes feeling unwell, I understand?'

'I was with him when he took off for Stourhead at Dunkeswell, and he didn't complain of feeling ill. He was looking forward to flying Sunburst in Austria.'

Mr. Gilmartin, when asked by the Coroner about this incident, replied:

'I understand from talking to the other triplane pilot, that if a pilot passed out, the aircraft would go into a shallow dive. In all probability, I would discount this theory.'

Summing-Up

Following an adjournment, H.M. Coroner David Masters brought the Inquest to a close at 2.30 p.m. After consultation with eight jurors having heard all the available evidence, the Coroner in his summing up concluded that as the rudder was a primary control in the triplane, it was most likely that its failure through a broken top hinge was the cause of the accident. The direction of the verdict was that Robin Austin Bowes died as a result of accidental death due to failure of the top rudder hinge and the subsequent malfunction of the rudder. This was considered an appropriate verdict on balance of probability by an occurrence that couldn't be foreseen.

With all said and done

Stepping out of the cool of the Coroner's Office and into the warm afternoon sun, witnesses were confronted by a small group of media occupying the narrow pavement. Hilary Higson commended Robin's bravery in steering his doomed aeroplane away from where they'd been sitting. Angela, when asked by a television crew whether she was happy with the result of the Inquest, said that she was and that "he was a very brave pilot."

Neither Ernie nor Sheila were asked for their comments.

Chapter 11 – Into the Sunset

At the entrance to Stourhead gardens stands an obelisk: the fourteenth century High Cross transplanted from Bristol in 1765. Walk the one and a half-mile circular path and you will find, aside from the temples set around an ornamental lake, its shores edged with rhododendrons, beech and tulip trees – bridged at its narrow neck; and from which can be seen a magnificent Pantheon, such hidden treasures as a grotto with nymph and water god statues. From the grotto follow the wooded path beyond the Pantheon to the Temple of the Sun. Between the Pantheon and Temple hides a twee woodman's cottage that doubtless had never been home to any woodman. Even before you enter the gardens a tiny hamlet comprising of Inn – The Spread Eagle – its courtyard, the church of St. Peter, and the row of cottages leading to the entrance beholds the interest of all those who wander down the lane from the car park; such tranquil rural peace and beauty in so few acres. The hamlet in itself is harmonious – exclusive yet also welcoming.

Walk into the Inn, order your drink at the bar then turn to look at the etching of a signature in one of the windowpanes. The vandals on this occasion were David Niven and his wife Pamela; the year 1941, a time in which the Hollyood star was serving as a commanding officer of a Liaison Regiment based at Stourhead.

Niven's signature is not inappropriate in this setting. The hamlet is reminiscent of a film set; a filmmaker's vision of what England may have been – perhaps should be. And if that can be said, then perhaps it can also be said, or at least suggested, that Stourhead is an antiquarian Disneyland. Wood cutters cottages that never were; mock temples to mythical gods; obelisks transplanted from their original sites; all of it nothing but an illusion. Created not to live in, but to view; to paint; a living, growing picture designed for the artist's canvas, and upon which the walker can intrude and become the subject of the artist's eye for one brief, fleeting moment.

Walk into Stourton House on a guided tour and you might see one such canvas painted by the artist St George Hare. His subject: a three-quarter length portrait of a 21 year-old British Army subaltern set against the backdrop of Stourhead Gardens; behind the subject the word "Kismet" etched into the bark of a tree. It is a portrait of a beloved family member – Henry Colt Arthur Hoare, the much adored and only son of Sir Henry and Lady Alda Hoare. They knew him simply as "Harry".

Harry 'Colt' Hoare 1914

It was not uncommon in those times for gentry families to capture the likeness of their heirs through the skills of a famed artist of the period following "the war" or "the Kaiser's war" as it was then known. However, this portrait of a beloved son that never returned is particularly poignant.

In 1895, City banker Sir Henry and Lady Alda together with Harry – then a young boy – relocated to Stourhead from their family home, Wavendon House in Buckinghamshire, initially occupying cottages on the estate while Stourton House was renovated following years of neglect by its former occupant, Sir Henry's profligate cousin, the 5th baronet.

As the Victorian age slipped away and the Edwardian era dawned, the costs of maintaining such a large property began to spiral. Parts of the estate had already been auctioned to pay off the 5th baronet's debts, but thanks to Sir Henry's intervention the essential core of the estate remained intact, though the high cost of running the mansion, gardens and wider estate were to remain a heavy financial burden for the new occupants.

"This Short Sketch into the Life of our Son"[4]

Lady Alda's bond with her only child had always been exceptionally close, their relationship being perhaps quite uncommon in an age of manners where children were often preferred "seen and not heard".

The family enjoyed a shared love of music. Intimate evenings would often be spent performing for each other in the Music Room with Lady Alda playing piano, Harry singing while Sir Henry enjoyed every moment. And when not ending

---

[4] Lady Alda Hoare's manuscript title, circa 1917-18

the day with such entertainment, Lady Alda and Harry would talk of the songs they loved and of song sheets still to be acquired.

Like so many of her contemporaries in the upper classes, Alda Weston had been musically trained from an early age, as an accomplishment in music was considered essential for courtship where the ability to entertain was crucial in finding an eligible spouse. To her credit, Lady Alda – a talented amateur musician – had a natural ear for music. Her performances on the piano became central to their home life at Stourhead.

Lady Alda Hoare

Despite the remote setting of Stourton House, life for the 6th baronet and his family was not entirely insular as like many of their class social soirees were the life blood of high society with the great and the good arriving by either carriage or motor car to drink, dine, gossip and dance. Author Thomas Hardy was a frequent guest and became a good friend: his signed edition novels depicting agrarian life in 19th Century Wessex lined the mansion's library shelves.

And so life might have continued had it not been for the German invasion of Belgium and, as a consequence of that hostile act, Prime Minister Asquith's declaration of war against Germany and the Austro-Hungarian Empire late in the evening of August 4th 1914.

Harry immediately enlisted in the local Dorset Yeomanry – his father's estate straddling the tri-county boundaries of Wiltshire, Somerset and Dorset. With him, the young men on the staff at Stourhead also enlisted and in so doing depleted the ranks of the workforce employed in the woods, grounds and the house until there were little more than a handful of older employees with some having to double up on their duties. Horses, too, were rounded up and Sir Henry was requested under emergency instructions signed by local magistrates to commandeer suitable mounts for the Yeomanry.

"Their Stourhead Mother"

As the war intensified and lengthened, it became clear to those waiting anxiously on the Home Front, that there would be no quick victory. Large numbers of casualties were returning home and the call for makeshift hospitals and respite centres went out across Britain. Volunteers were needed to offer their labour and practical skills for farm work or nursing. The landed gentry had the most to offer with mansions that could accommodate the wounded and grounds with outbuildings in which recuperating Tommies could convalesce – often before returning to the Front.

Provision was made at Stourton House for Tommies of all ranks and from all backgrounds to not only enjoy the house but also the gardens, with Lady Alda engaging them in conversation and entertaining them with her songs. She even encouraged them to visit whenever they so pleased, telling them that they were welcome to enter the house and to make themselves at home by the fire even if she wasn't always available to attend to their needs in person. The expansive grounds with their ornamental gardens were also open for them to enjoy: they could fish in the lake, play games or simply wander in the peace and quiet. And when not playing host to "her boys" as she called them, Lady Alda would visit the more infirm soldiers at the nearby military hospital in Mere. To many of them who came to know her, she was their "Stourhead Mother".

Lady Alda had a unique ability to empathise with the Tommies, and also with their mothers who, though far away from their boys, would – like her – be anxiously awaiting news from the Front. When Harry was sent home to England from Egypt in December 1916 to convalesce as recurring heart problems, double pneumonia, paratyphoid and exhaustion had overwhelmed him, it seemed a blessed relief that hopefully *his* war would at last be over.

Initial treatment had brought him home to Endsleigh Palace Hospital for Officers in London for convalescence and after six weeks he was sent home to Stourhead to the blessed relief of his parents. So for a brief time at least, a return to "normality" was enjoyed as once again the music room resonated to the sounds of Lady Alda's piano playing and Harry's rich baritone voice.

She wrote of this time:

"After dinner, playing at Harry's wish and he sang. While I fought for strength & tearlessness & found both. But, oh, the pain."

There would be just one more opportunity to sing again in that summer of 1917. Following his convalescence Harry was ordered to report to Curragh Camp in Ireland, having been declared "fit for light duty" by the Medical Board at Fovamp

Camp. A return to his regiment in Egypt was approved, but in the interim he was free to enjoy what remained of his military leave and so on July 7, he returned to Stourhead to be with his parents. Ten days later on July 17, he departed Stourhead for the last time en-route for the Middle East.

Lady Alda wrote of their final night together that she, Henry, and Harry spent: *"after dinner till past 10 in Music Room. Harry kept me playing his old favourites to him. Also he sang (beautifully)"*.

"One of the very last songs Harry sung me; on his last night at home, Sunday July 1917—and ever I hoped, the underlined words, would be prophetic (and often thought of them). But they were not— He never came back".

There is no day without its touch of sadness,
There is no heart without its need of care;
Still every *day has yet a gleam of gladness*,
And ev'ry heart a ray of sunshine rare.
After the clouds the sun,
After the heat the dew!
After the day with its cares grim and grey,
There's you, dear love, Always you!

Though dark the years with trouble and with sorrow.
The star of hope forever shines for me;
With you beside, I care not for the morrow,
For love is mine unto eternity!
After the clouds the sun,
After the heat the dew!
the day with its care grim and grey,
There's you, dear love Always you!
There's you, dear love Always you![5]

---

5 Italicized words indicate those that Lady Alda underlined in her score.

"Farewell at Morn"

Harry Hoare died of multiple bullet wounds inflicted in combat against Turkish forces at Mughair Ridge, Palestine in mid November 1917. Of the six wounds inflicted, one bullet pierced his lungs. Treated at the Ras el-Tin Military Hospital in Alexandria, Egypt, he lived a further six weeks and for a while he seemed to rally, but his heart condition, that had plagued him since childhood and had led to his earlier convalescence, contributed to his demise on December 19th, 1917. The tragic news eventually reached his distraught parents at midday on Christmas Eve.

Harry was, from the accounts of his fellow officers and those ranks that served under him, a popular and well-liked figure, highly competent and diligent in all his duties. Following his death, a friend wrote to Sir Henry and Lady Alda and described him thus:

> *" ... of unbounded spirits and energy he combined great pluck, tenacity and endurance with great keeness to learn all about his work, so as to be fully competent, ... He was most popular with all ranks. In Gallipoli, too, he had carried on, against Doctor's orders, till he absolutely dropped at his post from disease and exhaustion. I shall miss him personally very much. ... Many of his men have written letters home, speaking of his kindness to them ... He was one of the hardest workers and greatest 'tryers' I ever met. ... (He was) most anxious you should not know his condition and to spare you anxiety."*

This was a cherished letter that Lady Alda was to keep as part of her private memorial to her beloved son. She would return to "his" room, which since that fateful day had remained undisturbed, and re-read the letter every year on those dates that commemorated his last stay with them at Stourton and particularly on the anniversary of his passing.

"Alda - My Soldiers 1914 - 1918"

As part of her mourning, in the aftermath of Harry's death, a grieving Lady Alda set herself the task of documenting those last performances in her "Short Sketch" as she titled it, and also on her sheet music, annotating her memories and feelings in the margins. In a box of mementoes she entitled: "Harry box" she placed a wartime diary later adding two notes dated 1943 and 1944 stating that it should be purposefully preserved as part of remembrance; yet all her other papers she destroyed.

Despite her immense grief, Lady Alda continued to help her "boys" - the Tommies – those that were able to visit the house and grounds as her guests and also at the military hospital in Mere for those that couldn't make the short journey. All ranks from the highest born officer class to the lowliest private were treated equally and with equanimity as they joined in her songs, some of which had been Harry's favourite songs. She identified with them, her attention for their suffering did much to alleviate her own personal sorrow; she would play and sing popular anthems of the day, marching songs, love songs and laments encouraging them to join in or simply to listen.

She had established this connection with them before Harry's death and now she returned to the same format in a determined effort to fill the void. The songs were sung at afternoon tea or musical evenings for those convalescing or when visiting the Military hospital at Mere for those who were bedridden. She cherished their postcards, letters, and photographs compiling them in a special album entitled "My Soldiers, 1914–1918."

Of her heartfelt concern for the "boys" who were to be returned to the Front, she wrote in her diary:

"4 of my tonight 5 (Tommies) were going 'out' again—Oh! It wrings my heart—I play their songs—I keep outwardly gay, gay, gay—and I think of Harry—and I think of them, poor brave fellows; & of their

mothers—& oh! my heart it is torn at—it aches. Little Private Taylor spoke so nicely of 'my kindness to him, to them all'—Kindness?? Poor fellow . . . poor fellows—God knows its little enough to do for them. (God have mercy on my son—on all my soldiers.)"

On July 17, 1923, the seventh anniversary of his leaving Stourhead, Lady Alda visited the rooms that had been Harry's with all his possessions untouched and still in place exactly as she had ordered. Re-enacting the very same routine of that fateful night, she touched, held and inhaled those objects *he'd* held dear. She kissed his portraits and touched the mantle clock that she had stopped to mark the hour of his departure.

To mark the exact time he left seven years before, at exactly five minutes to three she wrote: *"waited, under [the] Portico, watching the old avenue, to Gateway, down which he then vanished for ever—then, the Hall-clock struck three, as I re-entered (just as that day it did, seven years ago, seven years ago) & my vigil was over. Shall I see many more?? God knows."*

At 8:15 she visited his bedroom. There: *"on the oak writing table ... his War-books as he* [had] *laid them down forever. ...* [his] *church-bells, beside them, his slippers, ...* [his] *ivory-brushes & toilet things , ... always his rooms look, still as were he but just gone—as if he must he must come back."*

> And *during all the live long day[s],*
> *Thy last farewell rings in my ear,*
> *In my forsaken soul holds sway*
> *The mystic sounds I seem to hear!*
> *Those farewell words upon me grate,*
> *Yet in my heart's recesses stay,*
> *Like crystal bells which vibrate*
> *Long after we have ceased to play* [6]

---

[6] "Farewell at Morn" by Emile Pessard, another selection performed by Lady Alda and Harry on his final night at Stourhead.

Harry was laid to rest in what is today the War Memorial Cemetery, Alexandria (Hadra), Egypt. Unable to have his body repatriated, in accordance with government policy not to repatriate the bodies of the fallen, Lady Alda mourned every day for the remainder of her life. Throughout the 1920s' and 30s', younger relatives came to know her as being "old fashioned" and "quite stuck in the era of the First World War," as she never attempted to update her hair or costume styles preferring always to remain as Harry and her Tommies had known her.

In 1938, as the shadow of yet another world war loomed on the horizon, a large part of the Stourhead estate which, in its totality covered some 2,650 acres including the mansion and ornamental gardens, was signed over to the National Trust by Sir Henry while the remainder passed to a cousin, Peter William Hoare with the formalities of transference eventually concluding at war's end in 1945. Two years later on the 25th March Sir Henry passed away followed just six hours later by Lady Alda.

They had stayed at Stourhead following Harry's death at the behest of Lady Alda though Sir Henry had longed to leave the place behind. *"I cannot bear to leave my son I everywhere see here"*, she wrote in her diary.[7]

Sir Henry wrote of his late, lamented son: *"Our only and the best of sons. He never grieved us by thought, word or deed."*[8]

---

7   'Stourhead Annals'.
8   Ibid.

Harry 'Colt' Hoare

On the 25th of August 1997, the day Robin would have celebrated his 53rd birthday, Wendy, her sister Angela with husband, Barry, together with Harry and Violet, and Christine, walked through the tranquil gardens toward Turner's Paddock. There they would wait for the small plane that would fly overhead to scatter his ashes. They had thought that morning that it would all have to be abandoned, as the weather was so bad – rain and overcast – horribly overcast. And yet now there was this beautiful day that seemed to be emerging from Heaven itself.

Standing together in the paddock, their walk over, the climb along the gently undulating wooded paths through those beautiful gardens done, they waited and listened, each with their thoughts; their silent prayers.

It was the strange inexplicable coincidences since his death that was on all their minds.

Angela: 'I couldn't believe it. I sat in Robin's car to drive it home and I put the radio on and what should be playing? *Those Magnificent Men in their Flying Machines*. I knew he was there, that he'd never leave me.'

Christine: 'Our arguments – always arguing as brothers and sisters do – him chasing me upstairs, grabbing my wrists. He had really, really strong hands. My wrists would bruise from his tight grip. He wasn't doing it to bruise me it was all just machismo. On the day of the funeral, when the coffin was being taken into the church, I moved to see what was happening and I leant against the pillar. The next day I had the most enormous bruise, which lasted for weeks. I can't answer that one, I really can't. It could have been coincidence, but I don't know.'

He was still such a presence for all of them; he still seemed to be a part of their lives.

Angela remembered the time she and Barry drove across the moors and another car pulled out in front of theirs.

'Barry went to swerve. The man saw us and he went to go the other way. Then he swerved back into our path again. We passed with barely a hair's breadth between us. Safely past we stopped and looked at one another. We would have been killed. Nothing left of us – outright. We both said: "Christ, that's Robin saved us!" We knew Robin had saved us, even Barry agreed. And Barry's very cynical, down to earth, no show, nothing. What you see is what you get with Barry. Robin saved us. We just knew he had.'

Wendy: 'These incredible dreams, it's always when I'm particularly low; such real dreams. I don't want to get up. It's not fantasy dreams. Sometimes in the dreams we argue – very real, ever so real. It's not nasty; it's like friendly sparring. But it always ends ... It's always nice. Now maybe that's my deep subconscious wishing it to be that way, I don't know.

'I still do not accept in my heart they don't know what happened. They have to justify their existence. Whatever happened it was an act of God. Robin did what Robin wanted to do in the nicest way. He did it. Even if it hurt him he would do it.'

Violet: 'That overcoat he bought from Burton's he kept like new until it got soiled with grease. Then he put it in a washing machine in the Launderette – an overcoat of all things – and it came out looking like new! He would turn his hand to anything. Me, with my experience of washing clothes and washing machines, I wouldn't have known. I'd have sent it to the cleaners. He just puts it in the machine and it came out looking like new! He'd have a go at anything. He was a good lad, good lad ...'

Harry: 'I can still see him walking with us to mum's house clutching that model car I bought him after we got back to London from Malta. Clutching it so tightly as if to say, "I must look after this."'

Violet: 'I think to be honest, if he lived with me all the time he might never have flown because I would have been on the safe side, especially with the small planes. I might have said, "Oh no, Robin," a mum's thoughts really. How nice it was of that lady who wrote after the accident saying she'd never known a kinder man. Robin was the most kind man she'd ever met, she said. So, it makes you – Because being away those years, we missed an awful lot really and truly.'

In the distance they could hear the sound of a small plane approaching. It seemed to be the only noise in the world that day.

They stood stock still, watching the sky. Then, all of a sudden, there it was, just above the treetops – the little plane flown by Dave Silsbury, and sat alongside him, Robin's nephew, Andrew, holding the urn that contained his uncle's ashes. Now the time to say goodbye was upon them. Dave banked the plane allowing Andrew to empty the urn. But it was from below that the most incredible vision could be seen, as the ashes gently descended to earth they were illuminated in the rays of the emerging sun whereupon they took on the aspect of gold dust falling.

No one noticed the plane fly away into the distance. As the ashes settled in the lush grass of the paddock, each of the family members turned to one another and hugged and cried. It was done.

One Last Tribute

When the little plane landed at the airstrip, Andrew took from his pocket a card given to him by his mother Christine just before he'd left her that morning at the Spread Eagle Inn.

'Carry it with you,' she'd said, 'and if possible, read it.'

He began to read it quietly to himself.

'What's that?' asked Dave, still emotionally stirred. Andrew showed him the card to which was clipped one of the last photos taken of Robin, alongside the now familiar poem that had been one of Robin's favourites.

'Please read it aloud.'

'Oh, I have slipped the surly bonds of Earth

And danced the skies on laughter-silvered wings ...'

Above My Head, A Distant Sky

After darkness had fallen on the second evening after Robin's death, Sheila sat on the patio step of the Old Inn house; at last the phone had stopped ringing. And it had been ringing constantly since the announcement of the accident. Mostly it was friends saying how shocked they were, how sorry, and what a wonderful man he was. There were calls too from Journalists; many of them had known Robin, or at least, had known *of* him.

It was pitch dark now. There were a few stars in the sky, not an awful lot; but where she'd seen the sun set, she saw two lights which, she felt, could not have been fixed wing aeroplanes as they were too erratic in their movements. The only thing they could possibly have been were two helicopters. Even so, they were *too close together and just too erratic for anything normally flying*, she thought.

She sat, mesmerised watching the lights. They were quite a way up in the direction of the moors. One came up to the other which appeared to be fairly static and moved around it – quite jerky movements; it jiggled around in this manner for two or three minutes before moving away to the east on its own. Then the one that had been reasonably static started moving up and up and up to the North-west whereupon it disappeared. The second light then continued right across the sky and went way down towards the South-east taking its time, moving slowly, but even then, bobbing around. She wondered again that the lights might be helicopters. They were not an unusual sight over the moors if an army exercise was under way. But it just seemed wrong as there was no sound, and usually in the quiet of an evening you could pick up the sound of all types of aeroplanes in the distance. And to add to the mystery, there were no flashing navigation lights, red or green or anything else besides, just this white light jiggling quietly across the night sky.

As she sat there watching the one remaining light eventually fade from view, she wondered that the scene she had just witnessed was something of a visual metaphor, the meeting of two spirits: the static light her husband, Howard, and the other light Robin making his way towards his old friend and greeting him:

'Hi Howard!'

And Howard looking at him and saying: 'Bosey! What the heck are you doing here? You're not supposed to be here.'

'Ah well, it's a long story. But I've been flying. I'm *going* flying.'

Epilogue:

He had started off as a car salesman to pay the rent and be near cars, but he soon became the complete flyer, the complete display pilot – the only one at that time to earn his entire living from it.

Although he cut short his own apprenticeship at Hawker Siddeley, no one that ever knew Robin ever said that he could no longer stand something any more. Robin seemed to enjoy everything he did and each stage was just that – a stage, a progression onto the next thing. Car sales progressed onto display flying; stock car racing progressed onto concours d'elegance. There was never any "good riddance" to a former past time.

Like cumulus clouds sweeping across a blue sky it was just one cloud following another. His motivation was creativity. It wasn't enough for Robin to fly, he had to create a picture story in the air; otherwise it meant little to him and the crowds below. His flying displays were his sketches come to life.

At day's end and in salute to Robin the flypast forms a "one man missing" formation at Rendcomb

Memories

The following tributes are from Robin's many friends and family members including: Roger Wilkin, Wendy Bowes, Christine Gillet, Gordon Clarke, Jeff Salter, Nigel Hamlin-Wright and Alan Jeffery.

"In later years we had played together. He came and spent a weekend up where I was at West Rainham in Norfolk, and we set up our gear in the mess and there was another chum of mine who lived in the mess there who used to play guitar; this must have been 1965?

"So there was three of us playing in the bar on a Saturday night, and the bar was the only place to be in the air force in those days 'cos drink was cheap; none of the drink/drive business, you could have as much as you liked and if you lived in the mess, you didn't have to go far to bed. And so Robin spent a weekend up there with us playing music.

"Latterly we had played at Northwood with a jazz band that I had played with in Lincolnshire, but we'd booked them for a jazz night and they were short of a drummer. I said, 'Don't worry about the drummer. My chum – he'll come and play drums.' And that's all it needed was a phone call to Robin and he rearranged his whole schedule around coming to Northwood from Plymouth – from Ivybridge – to come and play with us.

"The leader of the band, my good old mate Andy Hogg, was a little bit concerned, saying, 'Will your chum know what we do?' And I had to reassure him on a couple of occasions saying, 'Don't even think about it. After one bar you will realise that Robin will fit in with what we are doing. There will be no difficulty. And all you've got to do is look at him, raise an eyebrow and he'll do a drum solo. All you've got to do is look at him as you come towards the end of a number and just give him the nod and he'll give the two or three bars at the end. And don't worry about it, he'll be there.'

"After the first number, Andy said, 'I see what you mean. He's a bit good really isn't he?' And Sheila was there that night to witness that and we've got some photographs of that buried in the bookshelves somewhere. And that was a very, very successful night. He was a tremendously talented drummer." *Roger Wilkin, school friend from Malta.*

"Robin, together with Ken Hines, did attempt a John O'Groats to Lands End Driving record only to be foiled – as the story goes – by a stag charging across in front of the car causing a swerve up a bank and into a telegraph pole. But Robin's cups, trophies and medals are testimony to his many successes with stock-car racing, hill climb events and for the pristine condition of his lovely Aston Martin." *Christine Gillett, Robin's sister.*

Drivers and sales colleagues Ken Hines and Robin Bowes attempt to achieve a record non-stop drive between John O'Groats and Lands End

"I did a tour with the air force in Aden until it closed. I'd been to Turkey; I'd been to Cyprus. Every time I came home, every time without fail I would always give old Rob a call. And more often than not he was away because he was on the display circuit somewhere, and I'd just leave my message on his answer phone. It was interesting; his answer phone had the music of *Those Magnificent Men in their Flying Machines*, which was an obvious thing to do, and perhaps may have sounded just a little bit coy but it was completely in keeping with his outlook on life. I'd just leave a message saying, 'Hey Robin, this is me. I'm back in the country for the next week, give us a call when you can.' And, true to form, absolutely constant, good old reliable Robin. Yeah, he would always ring back, and he would always open with exactly the same thing: 'HELLO ROGER! Robin here, ..' And then we'd get into the business and we were back current again.

"He'd say, 'Now, let's see, I've got a display at Rendcomb, ... Yeah, ... I'll probably make a day spare there, yeah, okay, can we stop over with you?' And he'd make time; he'd make time to actually come and see you; he'd make time for people – wonderful! *Roger Wilkin.*

"But we knew he wanted to fly. From his school days his bedroom was a-wash with model planes, some on strings from the ceiling complete with cotton wool smoke trails, re-enacting a wartime dog-fight." *Christine Gillett.*

"And because I was a pal of Robin's, I was well embraced, and I was really very well looked after; it goes along with the measure of the guy, making time for us all. And he made provision for me; he wasn't at all selfish." *Roger Wilkin.*

"When Robin first learnt to fly as a hobby he earned an

award that enabled him to gain special night flight training and one of his earliest planes was an Auster. His flying proper started when he lived in Plymouth with his wife Wendy and he was sponsored by TSW. Then via one aircraft or another – a Cessna, Curriewot – 'Aireymouse', an old-time Nimrod – perhaps not in that order, he acquired a part share with Pat Crawford in the Fokker triplane and before long his part-time hobby became serious flying for him." *Christine Gillett.*

Robin and Christine posing by Crunchie GIIG

"Many's the time I've been speaking to chums about him in the past, and it'll happen again in the future, where I've recalled a particular story or a particular occurrence, and I've always been able to look back and say, my old chum Robin and I have done this at some particular time ... He was there all the time, and there is nothing I will do in the future that I can't reflect upon and say, 'Oh, Robin and I did something along those lines way back in the sixties or the seventies.' You know, like a brother, I'm a very proud man. Very privileged chap really." *Roger Wilkin.*

"… I was with Robin, and it was Valentine's day … and nothing came for me. And my sister and my friend were taking the rise out of me saying, 'Oh well with Robin Bowes it'll be Pickford's delivery van delivering yours,' and this and that, and I kept waiting for flowers to come. Nothing came. Robin was coming round that night. By the time he came round, six or seven o'clock whatever, I opened the door and I expected to see him stood there with something. But there was Robin and I couldn't contain myself because they'd had cards and flowers and had been winding me up, and he couldn't understand it. And his words were: 'But I'm here, aren't I?' Sort of: *I'm here, what more do you want? I'm here I'm giving you me for the evening, I'm here.* That wasn't big headedness at all, that was genuine. You've got me, you've got my love, what do you want? It took me time to realise this." *Wendy Bowes, Robin's wife.*

Wendy and Robin Bowes with Triumph Spitfire

'As a family, we were not traditionally flying enthusiasts and were slightly baffled as to where his desire to fly was inherited. And because of that and the distances involved we were not able to take an active part in his flying life. We saw him as often as possible at air shows (the performances) but were not there at the sharp end, the hard-graft preparation and didn't experience with him the frustrations when things went wrong or the whoops of joy when they went right." *Christine Gillett.*

"If you're coming across Mutley Plain towards Plymouth city centre, there's actually three lanes of traffic and we were in the outside lane and Robin was driving a demonstrator, he was working for Evans & Cutler and it [was] a Triumph Dolomite. Now the silly thing is, they've got two lanes but as you go you can only get into one because you have the bus lane that comes out – we weren't married or anything – Robin would just sit there, so casual and as soon as those lights changed it was like a bullet out of a gun. And Robin being on the outside had the right line because the inside one would go into the bus lane. I'm in the passenger seat, and this bloke's in the bus lane, … and Robin won't give. Robin won't give. And you're courting; you're not engaged, not married. If you were married you might say: "Oh for Chrissakes!" And I can see myself now … but Robin not looking left nor right just absolutely relaxed and calm. And this bloke ran out of bus lane and my Robin would just carry on." *Wendy Bowes.*

"Despite his quiet nature he had a good sense of humour – he loved Norman Wisdom's antics and never missed a Goon Show – and when the occasion warranted he didn't mind dressing up, although we would get his typical *do we need to know that* frown when we recalled him as a jonquil in an early school play. Can you imagine Robin dressed as a daffodil? We have photos of him as a pirate, a tramp, the butler in another school play Pride & Prejudice." *Christine Gillett*

"If you're a believer, you have to accept things sometimes without asking "Why?" I mean, if I ask, why? Why did Robin die that day? Nobody can give the answer to that. Nobody can. For whatever reason it happened ... There's times not to ask why, but to accept and believe. And in accepting and believing you get a comfort." *Wendy Bowes.*

"Robin was performing in the triplane at Rochester – the year escapes me. My dear dad, who has since passed away – Robin's Uncle Bob and godfather – and my son and daughter were invited to sit in the plane. We were all as proud as punch in front of so many spectators to be 'fraternising' with whom we considered the star of the show. Dad was even more proud when Robin had to leave the plane for a short while and asked him to mind it. I'll never forget the smug smile on dad's face and the pleasure and laughter it gave us all. We remember Robin with much love and intense pride." *Jennifer Swan* (Robin's cousin).

With the triplane at Badminton Air Day, Gloucestershire 1988

"It was the summer of either '67 or '68, Robin was most amused at the sight of me walking down the garden of Idenden Cottages playing my first bongo drums and wearing the Snoopy tee-shirt and straw cowboy hat Aunt Vi and Uncle Harry had sent me from the US. I don't know who was laughing more me or Robin. And I have the picture to prove it. Good memories." *Keith* (Robin's cousin).

"Robin's love of music has come through the family – Dad and many uncles played various instruments in the light dance-band style. Robin started drumming with the Boys Brigade, then a skiffle group with school chums, then later supplemented his income by playing at a couple of clubs in Plymouth, hotels in Torquay and with the Russ Thomas band. In recent years he joined Cats Whisker." *Christine Gillett.*

"The Students" school rock group playing to an appreciative audience in Malta

"But he was a pig headed sod at times, he would carry the drum kit, he'd carry these things and sometimes he'd be in pain. … That was Robin, you know, a man of steel in many ways but like soft marshmallow underneath. Sometimes I don't think the mallow came through enough." *Wendy Bowes.*

"And the fun we had; it was a real fun time. You can imagine that it was such a humour this man had. He wasn't an out and out comedian, but he had some lovely little ways with him whereby he'd tell you something, which was almost straight, but then you realised it was a huge gag. And it was; it was very funny; he had a wonderful way of saying things. And he could make you do things just by asking you certain ways, and before you know it you're doing these things." *Gordon Clarke,* (film maker and Robin's friend*).*

"Towards the beginning of July, … late one Monday evening, I had a phone call from Robin. He was settled into his little room at Rendcomb shortly before his trip to Austria and had turned on the radio to hear a programme all about 100 Oxford Street. That is the old haunt of the Humphrey Littleton jazz club and I know his own memory will have flown back over the years to the happy hours we spent there and at the other venues with friends very close to us." *Christine Gillett.*

"Robin was always nice, he was always polite, he was never difficult; if you wanted him to do anything, the answer was always 'Yeah, sure, that's no problem. I can do that.' His patience with people at interviews and this sort of thing was so unusual … interviewed in front of the Fokker triplane after it crashed in Germany; how patient he was with the TV crews who were asking him questions and there was his beloved triplane all smashed up in the background. And he was patiently answering their questions and asking them if that was okay? Did they want more? And that was really the sort of guy he was." *Jeff Salter* (friend and airshow organiser).

"Dear Robin,

Please find enclosed your commentary notes. May I both thank and congratulate you for your display on Sunday last at Weston Park. I know that you said you would put on a good show, but I really think you excelled yourself. It is not often that people whom one gets to know over the phone turn out to be as nice as you and I feel privileged to be able to say I now know you.

"Everyone on the ground clapped and cheered at the end of your display and a lot of people have made a point of telling me how much they enjoyed the triplane. For myself, I loved every minute of it. I hope that we can keep in touch and I look forward to our continued association.

"I should like to meet up with you at Yeovilton, but in any event I shall look out for your display. Once again Robin, thank you very much for making our show the success that it undoubtedly was. I look forward to receiving your invoice in due course and remain yours most sincerely." *W.J.C.*, (Show Organiser).

"He was a gentleman, he had so much time for people, it was just uncanny. You know, his was a busy life – a very busy life, but he found time for anybody. Even in the musical side of his life he found time to teach people or to go along and talk to them. And it really was lovely. Yes, the world is worse off for the loss of this gentleman. He was a gentleman flyer, much loved by everyone. As a matter of fact, when I went to Rendcomb, for the memorial day, you know it was so moving to find so many people there who had nothing but nice things to say about this man." *Gordon Clarke.*

"… it was very easy to, sort of, bask in that adoration and the interest of the aeroplane (Aireymouse), and what Robin said to me, and this was his ethos, I think, is that it's the aeroplane that's the thing the people are interested in. It's not you, the pilot. That was something I think Robin remembered throughout his entire flying career." *Jeff Salter.*

"I first met Robin when the 504 project was just a figment of my imagination. But having decided then that he was the man to fly her, it was a sound choice. Robin provided invaluable help lending his Warner engine mount for us to copy and advising on various aspects of the design, which would make the 504 the practical cross-country machine that it has turned out to be. When things were not going according to plan, Robin always had time in his busy schedule to enthuse and encourage. On more than one occasion, he made the long journey from Devon to Suffolk to view the 504's progress. 'Keep smiling mate, we'll get there,' was his often-repeated phrase. And get there we did." *Nigel Hamlin-Wight,* (friend and fellow aviator).

"Robin Bowes was Robin Bowes the man. He was *not* the Red baron! He played that part magnificently; it gave pleasure to many people, but Robin Bowes was a man who was brilliant at so many things." *Wendy Bowes.*

Historical accuracy: Robin was keen that the role of Red Baron be portrayed as accurately as possible and this included not only fine detail for the triplane, but also his own costume. Here, he's wearing an authentic uniform of the period in addition to the Blue Max – the Pour le Merite as would have been worn by the Baron himself.

"... at the end of the evening, he just came down and he always used to say, 'Alright? (laid back) Okay? Good!' You know, with a sort of matter of fact expression in those days. 'Yeah, that's good. Yep, nice music. Enjoyed that.' And I would always say, Did you know all the numbers? 'Oh, yeah, yeah, played all those in my time. Yeah, I liked those. It was good, good.' And it was all very, very sort of matter of fact, and that meant he thoroughly enjoyed it, and he'd been having a wonderful evening." *Roger Wilkin.*

"I took a look at him and he had this enormous band aid stuck across the bridge of his nose, and I said, 'What on earth happened to you?' He laughed somewhat painfully because it was still quite sore. At Liverpool they'd got out of the aeroplane, he had gone to the baggage compartment immediately behind the seat of the pilot; he'd opened it to get something out of the back and it was hinged up, and the wind caught it while he was ferreting around inside and pulled this metal hatch down across the bridge of his nose. Gave him quite a nasty whack. He said he was bleeding profusely and everybody was running around trying to think what to do and eventually they managed to stop his nose bleeding and he flew the rest of the trip with this big band aid across his nose. And later on that evening in the pub where all the crews stay, the rest of the team from The Crunchie Flying Circus all walked into the bar and all five of them, as I recall, had gone off while Robin wasn't looking and came back in with large blue pieces of sticky plastic stuck across the bridges of their noses. They all lined up behind Robin where he couldn't see them. I tapped him on the shoulder and he turned round and saw the team standing there grinning at him with these big, blue noses on and he just fell about laughing. And they spent the entire evening in the bar dressed up with these blue plastic bits stuck on their noses. What people thought in that bar that evening, I'll never know, but they didn't half make a funny looking sight." *Jeff Salter.*

"It was at the International Model Engineering Exhibition held at Wembley one year while I was doing a stint on the Model Power Boat Association stand, when Robin and a friend of his arrived. He was looking for an Air Brush (a type of small paint spray gun). I pointed him towards a stand that may be able to help him. I mentioned at the time that I did not know he was still into model aircraft. 'It's not,' he said, 'it's for the full size one!'" *Jim Free* (Robin's cousin).

"The first thing I learnt was how well known and popular Robin was amongst his peers, in fact it made me proud to know him, such was the respect in which he was held." *Alan Jeffrey* (friend).

"Aviation is well known for 'line shooters', not always a derogatory term; at least there is an infectious enthusiasm to share. One could never accuse Robin of such a thing, as he left his considerable skill to speak for itself." *Alan Jeffrey.*

"Robin continued the test programme, with me as ballast in the front seat, after driving up from Devon just for one day's flying. Finally the CAA issued the Permit to Fly. Once again Robin turned up in Suffolk for the 504's launch party, on a blustery Easter Monday. His polished performance and endearing nature charmed us all, not least my father whose second trip in a 504 was on the sixtieth anniversary of his first." *Nigel Hamlin-Wight.*

"At the time he was very much involved in Motor Sport, and amongst other cars, campaigned a V12 E-type Jaguar, … and a Ferrari 206 Dino, … These cars are now priceless." *Alan Jeffrey.*

"I like to think that Robin had a bit of a soft spot for the 504. It is, without a doubt, a Gentleman's Aerial Carriage and, although he probably wouldn't have it, they were well suited. Robin will be sadly missed and long remembered for his part in the creation and success of G-ECKE." *Nigel Hamlin-Wight.*

"I well remember him laughing, as he described how the engine on the 'Tipsy Nipper' he was flying cut out on him when he was upside down over Modbury, and putting the diminutive beastie into a field – deadstick." *Alan Jeffrey.*

"I once spent a fascinating afternoon with him, while he showed me around his triplane, and taught me something of its characteristics. Having just grasped the basics of flying a modern nose-wheel aircraft, I was in amazement at the dedication and skill required to master such an unruly device. It had to be flown by the rudder constantly, with virtually no inherent stability. It had a very fast idle, and required to be blipped on the switches to taxi. You could not see over the nose on the ground, and had to weave to see where you were going. As a tail wheel aircraft, landings had to be precise, with a tendency to ground loop if you got it all wrong. As a student pilot, I was very impressed!" *Alan Jeffrey.*

"I have many times watched his flying displays, in the Pitts Special in particular, and been uplifted by the spirit inherent in the loops, rolls and spins he used to perform with such ease. Unfortunately, although I love aerobatics, they don't love me and my body has proved unsuited to such delights. I did learn enough to appreciate the beauty of flight and the expression of the joy of life in such abandoned aerial ballet." *Alan Jeffrey.*

And surely the last word should go to the man himself ...

"I regard it as a privilege to be able to fly and present this aircraft to a public that has probably only ever been aware of the type through pictures in reference books. I hope that they get at least some portion of the pleasure from it that I do. Whether displaying to a crowd of anything up to 90,000 people or flying home alone on a summer evening, the memory and pleasure for me will last forever." *Robin A. Bowes.*

A keen aviation correspondent, Robin loved contributing articles for magazines such as Wingspan, the editor of which contributed £1,500 in sponsorship for the 1987 season. (credit Wingspan Publishing)

The End

About the Author

Geoff Pridmore's long held ambition was to be a writer, but it wouldn't be until the late 1980s that he began writing articles for national press and magazines as a freelance contributor.

A keen aviation buff he was a witness to Robin Bowes fatal accident at Stourhead Gardens in 1995 just prior to beginning a degree course in Journalism Studies. A year later, with the kind permission and encouragement of Robin Bowes' family and friends, he began researching and compiling material with a view to writing a biography that would document the life of a highly respected civilian display pilot who had made his name flying as the "Red Baron".

To comprehensively research Robin's remarkable story, the author spent 18 years compiling notes and interviewing the pilot's family, friends and colleagues – a process that took him across Britain and Ireland whilst trying to find a literary agent and/or a publisher.

In 2013, having completed a final draft of the book, an acquaintance encouraged him to self publish the biography. The title: *Not the Red Baron* having been inspired by a Tori Amos album track also of that title.

However, in recent years and following the completion of a Master's degree in History, the author – having uncovered new information that he considered relevant to Bowes' story – has included it in this edition for the first time.

Geoff Pridmore is also the author of *Teach Yourself Journalism* and his first fiction novel, *The Reunion* was published in 2020. He is a member of The Society of Authors and also The Biographer's Club. He lives in West Cornwall with his wife where he writes and gives talks on Robin Bowes life and work to clubs, societies and at festivals throughout the year.

The Story behind "Not the Red Baron"

I never met Robin Bowes, yet I was a witness to his untimely demise at Stourhead Gardens on that warm summer evening of July 20th 1995. An aeroplane enthusiast all my life, I'd always loved the sight and sound of flying machines whether jets, turbo props, helicopters or piston engined and would look skyward as soon as I heard an aeroplane approach. Growing up, I'd wanted to be a pilot but lacked the aptitude for maths and physics. The closest I'd come was as an Air Cadet and much later as a member of the Royal Auxiliary Air Force. For a time during the early 90s' I'd travelled to Popham Airfield from my home in West Wiltshire in order to clean the aeroplanes of a flying school in return for some "free" lessons.

At that time I might have met Robin, but our paths didn't cross and I remained unaware of the Red Baron triplane until July 19th 1995 at Stourhead, when I saw the WW1 dogfight display that made up just a part of the magnificent celebrations that marked the National Trust's centenary. Watching the tight turns of the Se5a and triplane replicas, I marvelled at a sight that my grandfathers might have witnessed from their trenches some 75 years earlier. In fact, I was so enthralled by the display, that I persuaded my elderly parents to see it and so together we returned the following night.

Following the crash at Stourhead, a series of rather bizarre coincidences began for which I can proffer no rational explanation other than to say that these occurrences led me to researching Robin's life story with the aim of completing his biography.

These coincidences and connections are far too numerous to list in this summary. On occasion they were quite bizarre; but it's enough to say that doors opened for me and research certainly became easier than it otherwise might have been.

It was only after attending Robin's inquest that I asked his family and friends about the possibility of writing a biography. The coincidences and connections had brought me so far.

As Robin's remarkable story began to unfold – told to me by those who knew him best, I realised just how naive I had been in my assumptions of people who fly aeroplanes for a living. I had thought, or rather simply assumed, that civilian display pilots were men and women of means with offshore bank accounts; that they were self-made or at birth fortunate enough to have inherited a fortune. How wrong I was!

Robin's background was similar in many ways to my own – no privileges, save for a post war upbringing, no special education, no family connections to call upon. He was an ordinary bloke earning an ordinary wage. In so many respects, Robin was the common man – yet he was also extremely uncommon. He used his meagre wages to afford an ambition that the common man might consider unobtainable. For example: his desire to fly, his desire to drive the ultimate in sports cars. "You can't do that!" someone might have said to him. Or, "people like us don't do things like that!" But I don't think anyone ever said anything like that to him.

He was clearly a man with a mindset – a positive outlook – that didn't prevent him from achieving his dreams; from achieving something that to others might have seemed unobtainable. He took the time to learn and practise whatever he felt he needed to achieve and so most of what he desired fell within his abilities. In addition to his loves of music, flying and driving, he learned to draw and stitch and sew and ski, so that by the end of his life he'd achieved all those things he'd set out to do.

I was sorry that I'd never met him, though simply to pay a casual visit in order to converse with Robin over a cup of tea in the kitchen, or to sit at a table for a meal, would have been to detain him from what he loved doing. To have known him would have been to work with him, to have kept pace with him; it wouldn't have been enough to drop round for idle chatter. That was not the way to "know" Robin.

This book is not only about the derring-do of Robin Bowes; it's about his remarkable friends and family, and they truly *are* remarkable. It is the purpose of this book to pay tribute to them too, and in equal measure. Their story is every bit as fascinating and as moving and I have tried to portray them as best as I possibly can. Every single one of them has welcomed me and been exceedingly generous with their time. Without their efforts and generosity in lending me valuable research material, this book would not have been possible. They have hid their tears when talking fondly of their dear, lost friend and I can only hope that I have listened and recorded their recollections as faithfully as I can. Should there be any omissions or mistakes in my transcription that for some reason has gone unchecked, I apologise most profoundly.

I must apologise too, in that this biography has centred largely upon Robin the pilot, rather than Robin the jazz drummer or Robin the motor sports competitor. And to those friends and colleagues who knew him in those fields – in which he brought so much experience and expertise – I am sincerely sorry that I have not been able to include more detail in respect of his remarkable life drumming and driving. To do so, would require another biography devoted entirely to Robin's passion for music and competitive motor sports, in which he undoubtedly excelled.

Nor is this book purely about machines – cars and/or aeroplanes – and those readers looking for a purely technical read might be disappointed. Throughout his career, Robin made copious technical notes relating to flying but to have included them all would have resulted in a hefty tome to have rivalled Tolstoy's *War and Peace*. I hope, however, that I've included enough information to satisfy their interests.

I hope, too, that my inclusion of other notable flyers and individuals such as Lt. Jo Kennedy, Percy Pilcher, D.H. Lawrence, et al, doesn't confuse the narrative too much, but these were connections that I found in my research and so I've included their stories along with that of the Red Baron himself.

Where I have been unable to contact either an associate of Robin's, or a witness to his crash, I have changed their name, as without permission I would be uncertain as to their approval for their name to be published. Such substitutions are indicated as a footnote. The names of officials previously published in the Coroner's Enquiry including that of the Coroner himself have been included.

I am indebted to those witnesses who have so kindly provided me with their testimonials relating to the fatal crash at Stourhead. Their lives were forever altered by the tragedy of that night of 20th July 1995 – an experience that I share with them.

I am also indebted to all those who have contributed material and in particular fellow authors Madeleine O'Rourke and Nigel Hamlin-Wright who have kindly given permission to draw upon their respective publications.

Harry Bowes and Gordon Clarke, had each filmed, recorded and archived much of Robin's flying, driving and various other pursuits on celluloid and video tape over the course of many years and without those archives my job would have been much more difficult.

Dave Silsbury and Jeff Salter both generously gave of their time by flying me in their own aircraft so that I could get a perspective of Robin's eye view from the air. These experiences were invaluable to me for gaining that perspective.

And thank you to the staff at the German Embassy, London, for putting me in touch with former ambassador Dr. Hermann von Richthofen, who kindly helped me to verify facts concerning his meeting with Robin at RNAS Yeovilton,

Somerset.

Robin's awards included: the Edgar Hamilton Trophy – awarded following the 1990 display season by the various air show organisers for the "Most Meritorious Air Show Performer".

Also: The Pooley Sword Trophy.

He was a Council Member of the Historic Aircraft Association and flew with the Great War Combat Team and also Vic Norman's Cadbury Crunchie sponsored Aerosuperbatics team.

This book is dedicated to all those wonderful people who fly to entertain.

Connect with the author

http://www.facebook.com/nottheredbaron

Maps

Map of Airshow venues p.142

Map of Stourhead display area p.224 (Credit: Air Show organiser for the Fete Champetre, Mike Smith

Printed in Great Britain
by Amazon